new bioethics

brave new bioethics

GREGORY E. PENCE

ROWMAN & LITTLEFIELD PUBLISHERS, INC.
Lanham • Boulder • New York • Oxford

ROWMAN & LITTLEFIELD PUBLISHERS, INC.

Published in the United States of America
by Rowman & Littlefield Publishers, Inc.
A Member of the Rowman & Littlefield Publishing Group
4720 Boston Way, Lanham, Maryland 20706
www.rowmanlittlefield.com

PO Box 317
Oxford
OX2 9RU, UK

British Library Cataloguing in Publication Information Available

Library of Congress Cataloging-in-Publication Data

Pence, Gregory E.
Brave new bioethics / Gregory E. Pence.
p. cm.
ISBN 0-7425-1436-6 (cloth : alk. paper)
ISBN 0-7425-1437-4 (pbk. : alk. paper)
1. Medical ethics. 2. Bioethics. I. Title.
R724 .P358 2002
174'.2—dc21

2002008962

Printed in the United States of America

∞™ The paper used in this publication meets the minimum requirements of American National Standard for Information Sciences—Permanence of Paper for Printed Library Materials, ANSI/NISO Z39.48-1992.

Contents

Assisted Reproduction

Cloning Embryos

Cloning Humans

Death and Dying

Money, Ethics, and Medicine

Ethics and AIDS

General Bioethics

Development of New Drugs

Food and Ethics

Preface

It would be dishonest to begin this book without sketching the events in my life that led to its conception. In 1974, I took a Ph.D. in philosophy from New York University and, after various other jobs, began teaching in 1977 at the University of Alabama Medical School in Birmingham.

Armed with classical theories of ethics and their subtlest variations, I intended to enlighten ignorant, southern medical students with the insights of a New York City–trained, analytic philosopher. It didn't work out that way.

One day about halfway through my first course on medical ethics, which was required for 165 second-year medical students, a student wearing a huge, hideous rubber head-mask snuck up behind me, placed his left palm behind my head, and smashed a pie into my face, knocking off my glasses and knocking me for a loop.

Seldom has a philosopher had so rude an awakening. I now understood the truth of the proposition that others didn't appreciate my discipline. I had given them the standard overview of utilitarianism, Kant, ancient Greek ethics, and allocation problems involving lifeboats and triage, but it hadn't worked. Three months into my first real job, I was a failure.

I also realized somewhat vaguely that the problem couldn't be just mine. Looking back, I see that the main problem was that medical ethics had only just begun and that no one really had much to teach. The final *Quinlan* right-to-die decision had just come down the year before, and it would be two more years before Louise Brown became the first in vitro–fertilized baby.

After getting the pie, I embarked on a journey to learn something to teach to medical students: I made rounds and saw patients undergoing hemodialysis, I spent every day for months with oncologists seeing cancer patients, I saw amniocentesis and medical genetics counselors at work, I witnessed surgery, and I met a lot of sick, old, scared people who just happened to be patients. Later, I spent five years helping to run the largest agency in our state for people with HIV infections, even becoming chair of its board of directors.

Eventually, medicine taught me more about how to do philosophy than philosophy enlightened me about how to do medicine. I can't practice medicine, so I settled for an improved version of philosophy. Not just any philosophy, but ethics, and not just any ethics, but ethics as it affects real human beings in medical settings at the sharp, human edges of conception, death, agony, and risk.

Later, and for many years, my pendulum swung too far away from philosophy: like beginning medical students, and most medical ethicists at some point in their careers, I became obsessed with getting the facts right. This was no small task, because neither in college nor in graduate school had I been especially well trained to deal with facts.

More recently, I realized that real philosophical issues always lurk in medicine and are always worth discussing; they just weren't the ones I had expected a priori and their resolution can't occur by applying some abstract theory or batch of principles. Like the proverbial phoenix rising from the ashes, real moral issues just *emerge* from the factual aspects of the case.

The essays that follow crystallize some of the things I have learned over the years. I was fortunate to have them appear in many leading newspapers and magazines, as well as Birmingham's major newspaper, with which I've had a happy relationship for more than a decade. The essays are patchworks, composed of bits from medicine, philosophy, recent history, literature, law, and every other perspective that illuminates what and how humans value.

Ethics can't be owned by any discipline because its subject matter is quintessentially human, and no academic discipline alone can capture the human condition. Ethics is about relationships between humans, and these relationships have financial, social, emotional, racial, sexual, as well as officially moral aspects. To really understand why moral dilemmas arise in facing death and creating life and, hence, to really understand how we ought to value and act, we have to grasp the full complexity of these relationships. For this reason, novels and biographies often capture the real ethical dilemmas of human life better than simple rules, equations, or slogans.

Somewhere along the line, I became a better teacher. In 1994, I won my university's highest teaching award. I choked when my name was announced because it meant—it seemed to me—that the pie episode was finally behind me.

Nevertheless, medical students, physicians, and the general public remain tough buyers for the philosopher selling his cognitive wares, and I don't always make a sale. What follows are some of my goods.

Acknowledgements

I thank philosophy editor Eve DeVaro and publisher Jon Sisk at Rowman & Littlefield Publishers for supporting this book. Julie Kirsch did her usual superb job at copyediting. Pooja Aggarwal, my 2002 summer research assistant, proofed the entire manuscript for errors; Jason Lott also found some. Thanks also to Alexee Deep for proofing the final book.

I especially thank a series of supportive op-ed page editors at the *Birmingham News*, starting with Terri Troncale (now with the *Times-Picayune* in New Orleans), Bob Blalock (now editorial page editor), and the current editor, Eddie Lard. Mary Arno of the *Los Angeles Times*, John Timpane of the *Philadelphia Inquirer*, and Jeanne Ferris of *The Chronicle of Higher Education* were also especially helpful.

1

Medical Ethics Is Whatever You Say It Is

In what seems like yesterday but was actually a quarter century ago, I stood in the office of the dean of medicine in the huge medical complex that exists in Birmingham, Alabama. I had just been told that I would henceforth have a job on both sides of the campus, a regular one in the Philosophy Department and a special one in the School of Medicine, teaching 165 students a required course in medical ethics once a year.

I was happy not because I had entered an exciting new area of interdisciplinary study but because I had any job at all connected to philosophy. During the previous year, I had come very close to having an alternate career in real estate in New York City.

After getting my Ph.D. in 1974 working under Raziel Abelson and Peter Singer (who visited for a year at NYU), I tried in vain for two years to secure a tenure-track job. Failing to do so, I had quit philosophy. (Looking back at statistics in those years, I now realize that the market was flooded with new Ph.D.s.)

Originally published as "Medical Ethics Is Whatever You Say It Is," *Newsletter on Medical Ethics of the American Philosophical Association* 1, no. 1 (Fall 2001): 164–65.

An unexpected opening brought me to Birmingham on January 1, 1976, when the Karen Quinlan case and trials were going on in New Jersey. The dean of medicine had studied philosophy as an undergraduate at Davidson College and, partly because of that and the Quinlan case, thought that medical students should think about such issues before they practiced. So he created a position, searched for an instructor, and ended up hiring me; the next fall I started to teach.

After he offered me the job, I accepted and then asked, "By the way, a lot of people disagree about medical ethics. What do you think it is?" He laughed and responded, "It doesn't matter what I think. From now on, it matters what *you* say. You're going to teach the course and write the books. Good luck!"

That was not the whole truth by any means, but when it came to teaching the course, it wasn't far off. There were no textbooks. My first year, I patched together readings from the *Hastings Center Report,* theology journals, *Journal of American Medical Association, Time,* and *Philosophy and Public Affairs.* Moreover, rather than embracing an exciting new area of interdisciplinary study, most medical students resented having to take another course (they didn't know when they had it good: since then, four additional courses have been piled on them: medical history, nutrition, neuroscience, and, soon, integrated problem solving).

During the early years, I longed for the safety of teaching traditional courses in philosophy—at least my courses in graduate school had some relation to that! I sometimes longed for my office on the green side of campus, far away from the hospitals where for several months I did daily rounds in oncology and other specialties. What was most uncomfortable was not having a definite role to play in making these rounds (for I was there strictly as an observer, not to advise about ethics; I have never been an ethics consultant in a hospital and, based on my experiences in hospitals, never want to be).

Until I taught in medical school and served on hospital committees, I had never realized the power a professor has in his mini-kingdom, his classroom, especially over the serfs who need a recommendation for medical or professional school. In a hospital or even a medical school, and especially in the early years, I often got the question, "What is a philosopher doing here?" I constantly felt I had to prove I belonged.

Over the years, I wrote a text (so I would have something to teach) and learned to like the constant challenges. Entirely unexpected issues kept arising, such as artificial hearts, AIDS, the Human Genome Project, and cloning. What I didn't like doing was trudging over to the medical library, or badgering physician friends, to try to quickly learn some new field.

I now think medical ethics is today's most exciting academic field. I was very lucky to enter on the ground floor and to help create a field that so well matched my interests and personality. (I am not constitutionally suited to spending decades writing on criteria of justified true belief or on proving to skeptics why I am not a brain in a vat.)

The growth in the number of courses in bioethics offered in North America has been explosive. (I purposefully use the wider term "bioethics" here rather than "medical ethics" to include issues such as genetically modified food and animal rights.) Each year, dozens of my medical and undergraduate students want to become bioethicists, and some have done so. Their interest is both satisfying and alarming—can the world accommodate so many bioethicists?

In September 2001, I attended the International Association of Bioethics meeting in London. What struck me was how bioethics has spread across the globe and is being taught in colleges in places such as Saudi Arabia and Liberia. The meetings also taught me that new issues arise when one takes an international perspective, such as the costs of AIDS drugs and the protection of human rights.

In 1976, the quintessential bioethical issue was death and dying. (There were also some red herrings: psychosurgery, genetic engineering, and behavior control.) What no one could have predicted were the cultural shocks from Louise Brown's in vitro fertilization and birth in 1978, from the failure of the *Roe v. Wade* ruling in 1973 to end the abortion debate, from the introduction of artificial hearts and xenografts (transplantable organs from other species), from the creation of hordes of homeless people due to deinstitutionalization in the 1980s, from surrogate mothers, from the discovery of genes for clinical disease, from thirty years of failure to provide medical coverage for the uninsured, and from the underground prejudice against gays and lesbians that surfaced and, with AIDS, continues.

Extrapolating from the last quarter century, I know that the big issues of the next twenty-five years will be unpredictable. Past experience would dictate that the big issues will be justice in medical coverage, AIDS, genetic discrimination, and expansion of parental choice about traits of children through new techniques in assisted reproduction.

But if the history of bioethics shows anything, it is that the biggest issues ambush the field. As Larry Altman, the physician-reporter of the *New York Times,* said in a retrospective in July 2001 (twenty years after the first public report of AIDS), every physician thought in 1981 that all infectious agents had been discovered. Altman wrote that at the time no one in medicine seriously considered the hypothesis that an entirely new, lethal agent had emerged.

Just as new infectious agents will emerge in medicine, so will entirely unexpected ethical issues. It is the nature of the beast. If you like to be challenged by unexpected issues (and to have reporters besieging you to take a stand almost instantly), bioethics is for you. Of course, you can always play it safe and inveigh against each new change, citing the mantra of bioethicists, the slippery slope. You can imply that each new way of creating families endangers the traditional family. You can be cautious and side with the American Medical Association (AMA) on most moral issues, but who needs philosopher-ethicists for that?

It is more interesting to take some risks and to back changes you think are reasonable. In any case, and as I wrote in *Re-Creating Medicine* (2000), medicine needs both "inside" and "outside" bioethicists to understand its problems internally and to critique them externally.

Personally, I believe that philosophers who understand the facts about medicine and science will play an increasingly important role in public policy and in daily medical life because no one else seems both willing and able to do so. If I am correct, then just as philosophy spawned the separate, new field of psychology a century ago, so one day we may look back and say the same about bioethics.

2

I Meet the AIDS Bigot

I have always felt like an impostor teaching medical ethics. Fifteen years ago, newly armed with a Ph.D. in philosophy from New York University, I never expected to be teaching medical ethics, much less in a medical school, much less about something like AIDS.

I didn't get the teaching job right away. The year 1974 was the peak of a bell curve for graduation of Ph.D.s in the humanities, and a vast oversupply resulted. After a few years, I grew frustrated, quit philosophy, decided to make big money, and sold real estate in New York City. There I encountered prejudice of all kinds. I made money, but to do so (I'm ashamed to say), I had to go along with the prejudice.

I later got a job teaching philosophy. I would like to say I gave up the big money of real estate because I hated the prejudice and loved philosophy, but this wouldn't be the truth. I couldn't accept failing to become what I had set out to be.

Originally published as "On the Road" (Guest Essay), *Journal of the American Medical Association* 260, no. 13 (October 7, 1988): 1946.

Although New York friends thought me crazy to leave, the job turned out to be better than they expected. During this time, Karen Quinlan's right-to-die case began, and courses in ethics were developed in medical schools, such as the one I was hired to teach. So it is that twelve years later, early on a warm April morning, I am driving on a two-hour trip to another city to talk on medical ethics. My host is a civic club that meets at lunch on a weekday and that often has a visiting guest speaker.

I am to talk about ethical issues concerning AIDS. I am not about to do this because I am a do-gooder. A local program of the National Endowment for the Humanities chose some humanities professors to give talks to nonacademic audiences, and this is mine. The idea is to bring the humanities to people, although sometimes this backfires. I have given this talk a half-dozen times before, and I guess I will give it a dozen times more, although I tire of discussing AIDS. As I crisscross airports and interstates with my carousel of slides (the security blanket of my medical colleagues), note cards, and evaluation forms, I think of myself as a mixture of peripatetic philosopher, sophist (I get paid), and dog and pony show.

When I arrive, I see about fifty middle-aged white men and ten women, clumped in small groups to gossip, a scattering of blacks among them. The young club president helps me set up the slide projector, then everyone has roast beef, potatoes, rolls, apple pie, and coffee.

Before my talk, we pray and pledge allegiance to the flag. New members are inducted, and a club officer explains the club is about friendship and civic duty. Most members seem to work in insurance, local banks, hospital administration, and the law. The president introduces a visiting female assistant district attorney. The city sees itself as progressive.

I feel nervous about this talk, even though it is only twenty minutes long. My problem is that my talk on ethics cannot avoid drug abuse and homosexuality, and I know these are hazardous topics. I soft-pedal the idea that most Americans are prejudiced against homosexuals. I say that, much as with racism and sexism of earlier years, people are still openly unashamed about prejudice to gays. I continue by discussing problems about blood banks, testing, and prevention.

When my talk is over, a powerful-looking man makes some grunting noises as he rises. His face contorts as he begins to speak. "What angers me," he says, "is spending my tax money on cures to save these faggots and drug addicts. I hope they just go away and die."

My face flushes and I stammer. The large man before me is a mass of rage and hate. The black members of the club look uncomfortable. I watch the big man speak as he continues to make even worse comments. It is impossible to stop him. I feel as if I'm paralyzed, watching a deadly force of nature erupt before me. I shuffle at the podium, feeling strangely surprised, angry, and incompetent. With the name of a well-known medical school behind me, I had expected the usual deferential audience, and here I am being attacked by a bigot. As he rants, several men snicker. I cannot tell whether they snicker with him or at the predicament of someone like me confronted with someone like him.

I debate a finesse, but finally say, "Your comments show the prejudice I am talking about." Unbowed, he instantly replies, "I'm not prejudiced. I just hate queers."

At this there are still snickers, but fewer. His comments are vile. This is an awkward scene. The arrogant, ugly face of prejudice has suddenly appeared; the audience's progressive view of their club and city has been challenged.

I keep on talking, mentioning that perhaps some people in this audience have gay children. ("Not true," the bigot mutters.) I say that past patients with cholera, syphilis, and plague were victims of similar prejudice. My statement falls on deaf ears. At this point, the younger members of the club rescue me, and for ten minutes I answer their (enlightened) questions. Afterward, a dozen members congratulate me and (in low voices) disparage the bigot. What they most want to tell me is, "Not everyone here feels that way."

Later, and like a lot of people in similar situations, I ask myself whether I could have answered the bigot better. It takes a long time to get home, so I have time to think. Should I have bothered with him at all? Was anything accomplished? It certainly made me feel bad. Certainly most speakers would have passed over the whole thing, pretending it didn't happen, just like when people say "nigger." I imagined myself with Richard Burton's voice and Dick Cavett's wit, refuting him while everyone laughs. It's a nice fantasy.

This confrontation could have happened in any similar club in the United States. Does it do any good to confront these things? Some medical ethicists are so moralistic about their opinions that they are counterproductive. It's certainly safer to play ostrich. I have colleagues who claim that confrontations like mine are never worth it, that they just polarize situations, that real bigots won't change, that on some level I did it to feel superior. Maybe.

Today, whites say that only a small minority of whites supported segregation and that most hated it. Yes, but how many confronted the bigots in our midst then? In New York or Chicago? In real estate, physicians' dining rooms, or civic clubs? I know I wasn't any big hero then.

My students sometimes say they wish they had lived twenty years ago and could have fought segregation because it was an evil without ambiguity and that the fight "merely" required courage. They seem to think there are no similar issues today. I'm not selling real estate anymore, so I'm ignorant about remaining prejudice in that area. But in medicine, AIDS has made one issue clear-cut. Maybe this isn't just a dog and pony show at all. Maybe I'm not an impostor. Maybe medical ethics really is important.

3

Do We Really Value Human Life?

I asked my students at LaGuardia Community College of CUNY in New York City whether or not a human life was worth at least $5. One student replied that the question was improper since no monetary value could be put on the priceless gift of life. Another student asked me to be more specific.

I then gave them a series of examples, all based on the assumption that real belief in putting a $5 value on life would lead to some kind of action.

Example one: suppose you are walking alone at 4 A.M. down a deserted street. A hit-and-run accident occurs. The victim, badly injured, is bleeding profusely. By chance, an off-duty ambulance passes. The driver is greedy. He says that the victim will surely die if not taken immediately to a hospital, and he will do so only if you fork over $5. Would you, assuming you have the money and it is no great hardship, give the man $5 to save a stranger's life, regardless of race, sex, or nationality?

Originally published as "The Value of a Life," *New York Times*, 4 December 1974, A23.

Almost everyone agreed that they would indeed give the sum. They were also unanimous in condemning the driver's selfish motives. One student asserted that he would only give the money to save a non-Caucasian.

Example two: The situation is the same, but suppose you are not there. Instead, your spouse, a trusted friend, or a relative telephones after witnessing the hit-and-run accident. They have no money with them, and they ask you to meet the driver at the hospital with $5. Some honest students admitted reluctance to get out of bed at 4 A.M. to save a life, but most agreed they would go if pressed.

Example three: if one human life is worth $5, are two lives worth $10? Are twenty worth $100? The logic was unassailable, but some students refused to go over $20. It began to appear to some that belief in the value of life is very shaky.

Then came the crucial premise of the argument: Can we really say a person has a moral belief if, given many chances to act on that belief, that person fails to act? Don't we call a person a hypocrite who says he believes something but when it comes to action does otherwise? The class unanimously agreed that real moral beliefs must be expressed in actions.

The conclusion: "You have now agreed to two premises: that human lives are worth $5, and that moral beliefs require action. I now accuse you all, and myself, of being hypocrites."

Why did I accuse them of being hypocrites? Because I believe that anyone who really cared about human lives in the last few years would have found some opportunity to give relief money to the starving populations of Bangladesh, Biafra, West Africa, and so on. There are, after all, organizations that help the needy using volunteer labor and with the vast majority of contributions going directly to relief efforts.

There were several objections. First, some students said that no one could know which organizations were really helping the starving. I suggested a few that did really help, and I asked whether now they were prepared to give. Only two students answered yes.

A second objection was that it was the government's obligation, not the individual's, to help the starving.

I pointed out the parallel to the Nazis' argument that the government alone, not individuals, was responsible for the mass killings of Jews. Governments are composed of people.

A rather disturbing, but enlightening, point kept arising in the subsequent discussion: whether a student's giving depended on the stranger's race, nationality, or religious beliefs.

Students were more likely to favor giving to help Americans than non-Americans. Jews were more likely to favor giving only to help Jews. Third-world students would not give to help starving Caucasians. (The student who at first said he would give money to the driver only if the victim was also black turned out to be very embarrassed by my conclusion that he should give to starving Africans.)

Perhaps this point tells us something of the nature of morality and how far we still have to go.

4

Exercise Is Dangerous to Your Health

I am running on my urban university's track, having momentary doubts about my religion. The track is next to Interstate 65, and I am on my fourth mile, my thirteenth lap. The day is hot, and I am tired.

I practice a secular religion of body and health whose orthodoxy decrees that exercise and preventive medicine will help me live vigorously to a hundred. A firm law of my universe declares that preventive efforts toward maximal health must work. This, after all, is the American Way: hard work, personal effort, and sweat must create a long, rich life. However, a recent incident involving spinach cracked the wall of my belief, fostering skepticism and agnosticism, which I here relate.

My spinach incident has a history. It started when I was eight, when one night at dinner I fulfilled some natural law of childhood: I refused to eat my spinach. Like all red-blooded American children, I knew that the gob of dark green, brackish material before me could not possibly be eaten. Like all such children, I eventually

---

Originally published as "Prevention Is So Much Spinach," My Turn, *Newsweek*, 25 August 1985, 10–11.

bowed to the powers-that-be and ate it. And like many adults, I learned to like this iron-rich vegetable, especially in fresh salads.

My fondness for spinach, however, helped precipitate a kidney stone, a most painful medical experience. Imagine my feelings of betrayal! Yes, my urologist said, spinach forms calcium oxalite crystals, along with cranberry juice (which everyone had also always said was good for me!), and I'd better avoid both, as well as calcium-rich foods such as milk and cheese. What about all those who had urged me to eat these items because they're "healthy"? Adelle Davis, Jane Brody, where were you when I needed you, when I was writhing on the hospital floor?

Sacrifice: Not that I have forsaken these priestesses of the Preventive Oracle. I still read everything I can about medicine and tune in to the health channel on cable TV. I tie up my running shoes each day and genuflect in the direction of the altar of sacrifice as I pursue the ideal muscle-fat ratio. At breakfast I'd rather eat salted ham and hash browns, but I down Cheerios instead. My diet includes more fiber and fish. I drink less than two drinks a day; I diet constantly. I've stopped smoking, and I wear my seat belt. Nothing can alter my belief that these activities will make me healthier and help me live longer: I am like those fanatics whose views cannot be falsified; I always have a counterexplanation. Prevention, you see, must work, or why am I going through all this denial and pain?

Still, one story haunts me. Many years ago I asked Kenneth Cooper, M.D., of aerobics fame, whether it would be better *not* to jog at all than to jog along smog-blanketed highways. He said it was a toss-up: that great damage could be done inhaling pollutants deeply during a three-mile run. Every runner I know runs along highways, especially at noon or in the evening, when auto pollution increases. I've told runners this and they all nod thoughtfully but continue to run near highways. Are we victims of the spinach fallacy? Will hundreds of thousands of us show up at clinics with impaired lungs?

Litany: Another pillar of my church maintains that prevention saves money. But when I add the costs of my preventive efforts to the charges for my recent hospitalization, I wonder where I (or the country) saved. Although insurance paid most of the bill, a family Blue Cross–Blue Shield policy now costs my employer (the biggest in Birmingham) about $2,400 a year, of which I pay $600. That's a considerable investment in prevention, on top of the costs for vitamins, health foods, health magazines, exercise clubs, and medical checkups.

Although "prevention works and saves money" continues to be the litany of my faith, I confess to doubts. During the last thirty years, we Americans have worked obsessively toward health, but if we're so smart, why ain't we rich? Why is the medical system now so expensive? Why is Medicare in trouble?

Obvious answer: we haven't spent enough on prevention! If we had, those elderly poor would be fit as fiddles today! But something doesn't fit here: the elderly population, although living longer, aren't living with fewer chronic diseases and aren't costing less for their medical care.

Nevertheless, people assure me that great savings from prevention are just around the corner. I know we saved a lot with dialysis and kidney transplants, and I guess we'll save even more with artificial hearts, for which many more candidates await. I just wish the savings would start showing up fast, because otherwise I'd have to not believe that prevention saves money. And if I did that I might be led to believe that exercise is dangerous to my health and that, as I see it, comes pretty close to blasphemy.

These thoughts scare me, so I'm going to leave off here because I have to see my orthopedist (I used to see ordinary physicians but I've been taught to see specialists directly, even though they cost more). You see, when I was running this morning, I stepped into a hole and hurt my ankle. I run now because I no longer play tennis—I have bursitis in my shoulder. Before that, I gave up yard work after my arm swelled up from a wasp sting. After I get my ankle X-rayed, I'll visit my psychotherapist so I can work through my new anxieties about preventive health, including the money I'm investing in seeing her. And at the rate I'm saving, I'm sure to die old and rich. Right?

5

Bioethicists and the Media: Finicky Lovers

Issues in bioethics are popular with the print and visual media. Such media have helped bioethics grow, and in turn, bioethicists have learned how to help reporters with their stories. A mutually advantageous relationship has developed, but within this relationship there are both strengths and dangers.

In general, as bioethicists have grown more skilled, they have become the perfect intermediaries between medicine and journalism. Physicians are by nature and training reluctant to talk to journalists; besides, if a physician seems too radical and too eccentric to his colleagues, his referrals may drop. Many bioethicists, with no patients and hence no referrals to worry about, are eager for attention from real-world journalists. The marriage is a natural.

When the Hastings Center in Westchester County, New York, began its work in the early 1970s as a completely private, self-sustaining institution, it needed the media's help to generate money to pay its staff and expenses. So the Hastings Center

---

Originally published as "Medicine, Media, and Bioethics," *Journal of Women's Health* 7, no. 10 (1998): 1217–23.

generated enormous publicity for bioethics in its first ten years, and its bioethicists became skilled in talking to journalists about bioethics.

Some professors and physicians will scoff that talking to journalists hardly requires skill, but they will be mistaken. Those who don't talk to the media much imagine that getting quoted is all about giving good sound bites. The reality is more complex. Getting quoted on television and in newspapers is mostly about forming a continuing relationship with a reporter. It is also about being willing to spend the time to sustain that relationship, sometimes flying to a distant city or taking calls all day instead of doing your own work.

If we survey bioethics over the last twenty-five years, we go from a time when Art Caplan was an intern at the Hastings Center to today, when he is the most well-known bioethicist in America and, perhaps, the world. It is hard to watch any episode of *60 Minutes* (one of the most-watched shows on American television) on any bioethics topic without seeing Caplan interviewed. In large part, this is because he is able to talk to the media in a way that everybody can understand, because he has the time to do so, and because he really wants to do so.

One great difference today versus thirty (or even twenty) years ago is that bioethicists themselves are players in defining the moral issues, bioethicists themselves are interviewed on television, and bioethicists themselves are outing moral issues. This new development has both advantages and disadvantages.

The danger of this relationship is that it is easiest for a reporter to call the same bioethicist over and over again. Moreover, it may be difficult for a bioethicist to suggest the reporter speak to other bioethicists who hold different beliefs than his or her own.

Certainly every American by now realizes that medicine is rife with moral issues. Such issues are routinely reported by the television news shows, national weekly newsmagazines, and newspapers (especially the *New York Times, Washington Post,* and *Los Angeles Times*).

When there is a bioethicist character on *ER*, or Barnes and Noble creates a section for bioethics books, the field will have been canonized.

The great danger at present is that only a very few media-savvy bioethicists relay to the public what bioethics says about an issue. Moreover, too often only one side is presented as "the" ethical position. The two best examples of this in the 1980s concerned surrogate mothers and physician-assisted dying. A few well-known bioethicists staunchly opposed these changes and were quoted over and over in the national media. Yet their positions failed to reflect the diversity of opinion on these issues among bioethicists, the public, and physicians. Indeed, when two-thirds of Americans favored liberalizing laws about physician-assisted dying (as Oregon recently did), two-thirds of physicians opposed doing so, and these bioethicists sided with the conservative physicians.

Such one-sidedness is especially dangerous in discussing reproductive ethics, where Americans burn red-hot with passion. So far, any issue even distantly related to the abortion debate is automatically inflamed: witness the recent criticisms of the use of stem cells derived from human embryos, where sanctity-of-life champions repeatedly call period-sized embryos "tiny babies."

Another problem about bioethics and the media is that journalists like to quote people on extreme positions, so that the journalist appears in the middle as a reasonable, unbiased person. This allows any extremist to skew the debate.

But consider abortion globally. In Romania under Ceaucescu, abortions were illegal, so pregnant women had to undergo *forced birth*. In China, after one child, abortions were mandated, so there was *forced abortion*. Obviously, the rational compromise is to let such decisions be up to the women involved.

Yet in America, the pro-choice position is not reported as a reasonable compromise but as an extreme stance. How on earth did our media arrive at this irrational position? Yet few bioethicists challenge the assumed parameters of this debate.

The great thing about the media today is that breaking issues in both medicine and medical ethics are reported with breathtaking speed. The downside is that, in the rush to be first, major mistakes are easily made.

Whether the media can accurately and quickly convey the complexities of breaking developments remains problematic. The recent reporting about issues of human cloning has not been distinguished by its quality, as many idiotic comments have been reported unchallenged by countering, sensible comments. Meanwhile, inside medicine, countless experimental uses of old drugs, traditional techniques, and formal experiments receive little or no attention because no one brokers such stories to the media. There is thus an asymmetry in coverage, where the official bioethics issues receive saturation coverage and the more typical, but no less profound, ethical issues of everyday medicine are not even discussed.

Bioethicists themselves bear some responsibility for some of the media's problems in covering issues of ethics and medicine. Rather than saying, "In my opinion, it's wrong that . . ." they too often aggressively assert, "IT'S WRONG that . . ." While this is a good trick for intimidating opponents in seminars, it comes across in public as arrogant. Moreover, because equally rational and sensitive bioethicists have come to different conclusions, it is unprofessional not to represent those opinions too.

Indeed, bioethicists have no special authority to speak about right and wrong. There is no committee or exam that only lets *moral* bioethicists get Ph.D.s or positions. There is no special sort of moral wisdom that bioethicists possess.

The condemnation or endorsement of a medical development by a bioethicist in itself carries no weight, nor should it. The only thing a bioethicist brings to the public arena is her own arguments and reasons. If she is good, she has good reasons and evidence for her position, and she should give such reasons cogently. If she is fair, she will acknowledge that there is opposition and reply to the best objections of that opposition.

Bioethicists should also beware speaking in platitudes and g
ing knee-jerk answers to reporters' questions. Is it an occupatic
disease that a bioethicist cannot go on television without menti
ing the slippery slope? A symptom of the same disease is to im
that each new option in human reproduction is an "assault" on
family and motherhood.

We bioethicists should especially learn to resist invoking
mantra that "medical technology is changing faster than our al
ity to think through the moral issues." This platitude implies t
some day we will achieve moral perfection in society, and then a
only then will we be able to judge new medical technologies c
rectly. Please! Is this what we really think?

The opposite is almost certainly true: that we learn and g
moral wisdom *precisely in judging social change*. If we didn't have so
change to test us—if an important, possible change didn't cause
to really think about whether it is desirable—we wouldn't get a
wiser. An important part of moral thought is deciding wh
changes in society are bad and which are good.

Bioethicists can bring one special thing to the public: kno
edge of the history of bioethics. Every issue in bioethics ha
pedigree, from those at the end of life to those at the beginni
From assisted dying to assisted reproduction, someone has thoug
about the issues before in some way. That in itself may be help
for many to know and may exert a calming effect on reporting
issues sensationalized by the media. Such information may coun
the impression that infanticide, medical research on humans, a
the view of society toward "sinful" patients (for example, AII
victims) are special problems of our age.

Finally, the hardest thing for a bioethicist to do is to decline
be interviewed because the reporter is presenting the wrong view
a story and won't change. A related difficulty is when the report
wants the bioethicist to say that the sky is falling when the truth
just the opposite. Since getting on television and being quoted l
famous newspapers helps the bioethicist to earn large speakin

fees, the temptation to say what the reporter wants is enormous. Nevertheless, there are frequent occasions when the ethical thing to do is to de-sensationalize the story, to substitute reasons for emotions, to counter prejudice with facts, and, in short, to be a professional and a scholar.

6

Re-Creating Bioethics

Scientists are producing vast amounts of new information about human biology. Questions about who should have access to that information, and how it should be used, make the field of bioethics increasingly important to a growing number of people. But bioethicists' work is often too simplistic, even sensational.

How most of us in the field think about the Human Genome Project is a telling example. Typically, we emphasize the dangers of discrimination by employers and insurers, who might misuse genetic information about their employees and insureds, and we view as villains both insurance companies and businesses that own patents on genes. However, the situation is far more complex.

For instance, businesses with patents will push for strong laws against some forms of discrimination. If people fear such discrimination, they won't take genetic tests and therefore won't buy the products that those businesses want to sell, such as medicines designed specifically for each patient's genome.

Originally published as "Setting a Common, Careful Policy for Bioethics," *Chronicle of Higher Education*, 12 January 2001, B20.

On the other hand, insurance companies will resist laws against some forms of discrimination. They want to be able to refuse to sell a million-dollar life insurance policy to someone who has just discovered he is infected with HIV. California once barred insurance companies from refusing to sell life insurance to residents of the state who refused to be tested for HIV, but then the companies refused to write new policies for anyone in the state, and the legislature repealed the law. Even life insurance companies deserve some sympathy—they could easily go bankrupt.

How can bioethicists help policymakers and the public understand the complexity and the importance of such issues? A good start would be to avoid the following mistakes, which are common in the current practice of bioethics.

Seeing complex problems in simplistic terms. Wittgenstein noted that while the first step often escapes notice, it sets up the conceptualization of the ensuing debate. Reducing multifaceted debates to two opposing, oversimplified views is often the wrong first step in discussions on bioethics. For example, it is simplistic to think that we can acquire organs for transplant only through altruistic donations or in a crassly commercial market. Although some people do not realize it, selling blood became legal in the United States decades ago, and voluntary donations have not decreased as a result. My point here is not that selling organs is a perfect solution but that the disadvantages are outweighed by the fact that four thousand Americans now die each year waiting for transplants.

Thinking in black-and-white terms makes it easy to keep discussing the same viewpoints, and hard to come up with new solutions.

Insisting on Olympian standards for new options. Bioethicists demand saintly motives of physicians and couples using new services and want guarantees of healthy babies. On the other hand, they ignore the everyday dangers to children born through traditional methods, like the millions of babies born to pregnant women who drink alcohol or smoke. Of course, we always have a chance to do better with new methods, but we shouldn't demand perfection in those cases when we set lower standards with traditional methods.

Distrusting the choices of ordinary people. When parents at risk for genetic disease test themselves or their embryos, they are accused of wanting only perfect babies. Bioethicists almost always equate parental desires for better children with eugenics. Yet new technologies give parents more choice. Real eugenics—in the United States in the early twentieth century and in Germany under the Nazis—tried to take away choice.

Bioethicists should also be consistent about choice. We shouldn't condemn choice when we discuss reproduction, but applaud it when we talk about people's right to die.

Confusing questions of access with questions of value. Before we worry about how to distribute something scarce, like lifesaving dialysis machines, we need to determine that the machines are good. Egalitarians say that we should not create smarter children through genetic enhancement or drugs because we cannot guarantee equal access to such options. But we should first determine the value of the option.

Demonizing new inventions. Bioethicists often see new technologies as threats to our humanity, rather than as expressions of it. In the 1970s, many bioethicists feared that allowing amniocentesis would lead to couples' aborting less-than-perfect babies. The critics failed to realize that parents would abort those desperately wanted babies only if the test revealed a devastating abnormality. In fact, new inventions are just neutral tools. How we use them determines whether their effects are good or bad.

Letting sensational cases skew our thinking. When one surrogate mother refuses to give her baby to the couple who hired her, or when a group of women allegedly wants to auction their eggs on the Internet, we incorrectly generalize to a thousand other cases of surrogacy or egg donation where nothing of the sort occurs. So blinded are we by such cases that we fail to heed the normal rules of evidence, reason, and arithmetic. Phrases such as "reproductive technology," "genetic engineering," "test-tube babies," and "clone" now have such sensationalistic connotations that they hamper rational debate.

Ignoring the fact that people often have mixed motives for their actions. For instance, people have children not only because they love kids, but also to add meaning to their lives, see part of themselves continue, and have company in old age. Public policies like those for patenting genes, donating organs, and making choices about medical treatment for children need to take such mixed motives into account.

Failing to note the cost of doing nothing. When we discover new ways of getting organs for transplant or helping people have children, bioethicists characteristically raise every possible objection. We often assume that it's better not to change. But the status quo is not ideal, and being too cautious to try new approaches also has disadvantages.

Ignoring the role of money. Some bioethicists demanded that African countries be allowed to make cheap drugs to fight AIDS, even if they violate international trade and patent laws in the process. But if everyone could do that, drug companies would go bankrupt and no new drugs would appear. Instead, bioethicists should look for practical solutions, like allowing patents on drugs but—in cases of great need—requiring the patent holders to license their products to other companies that will sell them more cheaply.

Failing to advocate positive changes. Too many bioethicists enjoy being naysayers: Physician-assisted dying is wrong, human cloning is unthinkable, paying surrogate mothers or egg donors exploits women, genetic therapy is too dangerous, and allowing couples to choose their children's traits leads to eugenics. We are trained to predict doom rather than to support progress.

But we used to be reformers. As bioethicist Nancy King, of the University of North Carolina at Chapel Hill, noted in a 1999 article in *HEC Forum,* "One of the original roles for people in bioethics was as the 'outsider,' the moral stranger, the one who tried to help others find their voices." Too many of us who work in hospitals or medical centers run the risk of simply ratifying physicians' beliefs. We must preserve our independence and our outsider status.

Bioethics is a vital discipline in our technological world. We need its practitioners to produce careful analyses of, not just knee-jerk reactions to, the many subtle problems that confront us.

7

Bush's Bioethics Council: Dead on Arrival?

President Bush announced at the start of 2002 the seventeen members of his new Council on Bioethics, chaired by University of Chicago bioethicist Leon Kass, the most reactionary bioethicist in North America. Is this council dead on arrival? Its members look to me to be a pretty conservative bunch, especially theologian Gilbert Meilaender, who espouses conservative bioethical views with Kass in *First Things*, and Francis Fukuyama, who wants to curtail biotechnology. James Wilson is a pal of Kass' and journalist Charles Krauthammer is one of Kass' biggest fans.

Can such a conservative group represent the best in American thinking about biotechnology? Perhaps, but I doubt it. Reasonable people believe that infertile married couples have the right to use in vitro fertilization to try to create kids, but not Kass, whose view echoes that of the Vatican. Kass says that the views of Orthodox Judaism resonate sympathetically within him, but his receptors need tuning, for most Orthodox rabbis consider assisted reproduction quite permissible to create a family.

Originally published as "Is President's New Bioethics Council Dead on Arrival?" *Birmingham Post-Herald*, 24 January 2002, A14.

White House spokesperson Ari Fleischer said, "The council will keep the President and our nation apprised of new developments and provide a forum for discussion and evaluation of these profound issues." How exactly will that occur when none of the members really thinks out of the box?

Take cloning. Are we going to hear both sides on embryonic and reproductive cloning? The House of Representatives has voted to go against the U.S. Constitution and make both a federal crime. Are we going to hear more rational views? To my knowledge, none of the members favors reproductive cloning. But how can a view be evaluated and discussed by such a group without a strong advocate? That's like going back to 1972, before *Roe v. Wade*, and appointing a council with no pro-choice members.

And who are these people? I recognize Bill May and Rebecca Dresser as working bioethicists with a lot of experience in this field, but the appointment of others mystifies me. Michael Sandel is a well-known liberal political theorist, and Stephen Carter is a law professor who writes about religion, as well as novels, but neither of them is a working bioethicist. That's important because good recommendations about brain death, transplants, stem cells, artificial wombs, insanity, dignified dying, gene therapy, and AIDS do not come from reading a few reports but from long study and experience.

What is the function of this Bioethics Council anyway? To express a consensus at the most minimal level? If so, remember that no such consensus existed before the first test-tube baby was born: Steptoe and Edwards went against popular fears and just did it. Similarly, no such consensus existed about legalizing abortion.

This is important for two reasons. First, if Kass and others still have not accepted in vitro fertilization or abortion twenty years after the rest of us, will they ever? Second, perhaps government shouldn't seek ethical consensus in a pluralistic society; bioethics concerns medical progress, an elitist enterprise. Medical innovation doesn't occur by polling citizens or a committee's final consensus.

In my view, the Bioethics Council should lead the nation, not anoint agreement when the last ignoramus finally comes around. To lead, of course, would mean taking big risks. That's how real progress occurs.

Pulitzer Prize–winning author and MacArthur genius–award recipient Jared Diamond argues that aliens surveying Earth in 1500 would have predicted that China would rule the world today, but that did not happen because China's emperor decided that technology had gone too far, too fast. In his study of why some societies stagnate and others flourish in *Guns, Germs, and Steel,* Diamond writes, "We tend to assume that useful technologies, once acquired, inevitably persist until superceded by better ones. In reality, technologies must be not only acquired but maintained, and that depends on many unpredictable factors."

One such factor for medical-scientific advance is enthusiastic support, not moral condemnation. No big leaps of innovation ever came from naysayers and cautious committees. Perhaps I'm wrong, but I don't expect many green lights from this group for bold research. For my money, that's a shame.

8

On Reading Shakespeare to Get into Medical School

Opinions about medical school resist facts. Nowhere is this more true than in advising premedical students about selection of college courses (the perennial "how-to-get-into-medical-school" question). The Rockefeller Foundation recently discovered the inaccuracy of most such advice, especially to nonscience majors. Well-meaning high school counselors advise college-bound students to take premed sequences (only science courses), so premed becomes a major—where no such major exists.

Premeds supposedly must declare allegiance to science at an early age, perhaps as early as junior high school, but certainly by mid-college. Premeds, parents of premeds, the general public, and even some physicians believe that more science makes for higher acceptance to medical schools. Not true. Similar myths connect courses in political science, criminal justice, and business law to admission to law school.

The best antidote against this pathological immunity to facts is evidence. Patterns of acceptances of students into medical

---

Originally published as "On Reading Shakespeare to Get into Medical School," *Alabama Journal of Medical Sciences* 22, no. 1 (January 1985): 19–20.

schools across the country show that, contrary to popular belief, those who major in chemistry and biology do not have the highest acceptance rates. Music (58.1% of music majors who apply to medical school are accepted), philosophy (55.8%), and economics (53.0%) surpass biology (42.6%) and chemistry (52.9%) (source: applicant files, Association of American Medical Colleges).

One might think that with such evidence, premeds would flock to these majors, enthusiastically embracing other fields, studying whatever subjects excited their strongest passions. Not so. Premed Jacks and Jills fear tumbles down the straight and narrow hill and refuse to follow evidence off the well-beaten, suffocating path left by previous premeds.

It is almost impossible to get premeds to believe these real data. The few social sciences and humanities students who do apply to medical school usually start out in fields such as English or history, then switch to medicine. Often such students graduate with Bachelor of Arts degrees and only later take physical science courses as irregular postgraduates.

"But even if this were true," premeds will chorus, "we'll be hurt competing with science majors on MCATs [Medical College Admissions Tests]." Where is the evidence? (Supposedly premeds learn to ask such questions in science courses.) Contrary to popular belief, science questions on MCATs come only from introductory courses in biology, chemistry, and physics.

Of course, nonscience majors need these first-year courses to enter medical school and take the MCATs, but they need not take further science courses to do well on the MCAT. The exam fundamentally tests reasoning, not recall—it is an aptitude exam, not an achievement exam like the Graduate Record Examination (GRE). Some of the highest MCAT scores come from nonscience majors with only first-year physical science courses. (What about Kaplan courses? Don't they raise MCAT scores? Yes, they do, but this doesn't disprove the claim that MCATs are aptitude tests; rather, it proves it. Kaplan courses merely organize materials for study on a national level to counter deficiencies in particular colleges and courses. They do not teach nine science courses in a few weeks.)

"But if this were true, won't I do worse in medical school when I have to compete with science majors?" Again, where is the evidence? Some medical professors claim undergraduate courses do not teach the basic material well enough, so they start all over anyway. Most nonscience majors do feel disadvantaged in some basic science courses (e.g., in anatomy), when many premeds have already studied anatomy. Nevertheless, nonscience majors quickly overcome such deficiencies. Of course, a medical student who had undergraduate anatomy will, with the same amount of work, be more likely to get an "A" in medical anatomy than the anthropology major without that background. That is to be expected: A major should count for something. But the big question is: Does it all quickly even out?

One answer is that some practicing physicians think medical schools spend too much time on anatomy and that it is ridiculous for undergraduates to study anatomy instead of Socrates and Beethoven. Another answer comes from the Rockefeller Foundation's conclusions: In a study at one New York state medical school, in fifteen basic science courses and scores on Part I of the National Board of Medical Examiners examination (NBME), there was "no significant difference in performance between science and nonscience majors." In fact, in eight of these fifteen courses, nonscience majors outscored science majors.

What about the clinical years? Again, the Rockefeller Commission, whose members included noted physicians such as Lewis Thomas and Edmund Pellegrino, concluded "there was no significant difference in clinical course grades or scores on Part II of national boards." Finally, no statistically significant relationship was discovered between undergraduate major and selection of medical specialty (humanities majors do not go only into "soft" specialties).

This brings us to the final, perhaps most important, question: How do undergraduate majors affect physicians' practices of medicine and their personal lives? The first part of an answer recognizes that medicine is still neither just science nor just technique: It is still "the healer's art" (to quote the title of Eric Cassell's excellent book). Good medicine intellectually involves memory and intelligence. These qualities are distributed evenly among science and nonscience majors. Good medicine also involves compassion, courage, integrity, resistance to drug dependency, fairness, friendship, and veracity. It is at least possible that discussion, reading, acting, and writing about topics such as war, love, death, and morality supply a good premedical background.

Perhaps more important than memory and intelligence is the nonscience major's reflection on the agonizing, complex issues he or she will face practicing medicine in the twenty-first century. Of course, it would be arrogant to suggest that physicians do not think about euthanasia, truth-telling, and greed, but not every physician has *systematically* thought about these issues, with the help of reading great thinkers, and under the guidance of an interested professor.

Indeed, it seems that there are three crucial times in a physician's life when he or she is likely to be open to such influences: as a premed student, as a resident, and in retirement. Medical schools have little contact with retirees, residents have no time for anything, and, hence, only premeds are susceptible to the novelist, musician, philosopher, and dramatist.

Medical schools should teach medicine; undergraduate colleges should teach traditional majors. One major is, in fact, as good as another for getting into medical school. If this advice were borne more in mind, we would see more educated physicians and fewer thirty-year-old premeds applying to medical school for the seventh time because they were educated for nothing else. Who knows? We might even see better physicians.

9

Happy Twentieth Birthday, Louise Brown

Twenty years ago today, Louise Brown was born through in vitro fertilization amid alarmist predictions that her birth would shake the foundations of human reproduction. Jeremiahs such as Jeremy Rifkin carped that babies such as Louise would be harmed by being rubbed (as embryos) against glass and steel. Scientist-turned-bioethicist Leon Kass warned that the family and society would be harmed by sundering the bonds between the act of sex and conception.

So great was this exaggerated alarm that Louise's parents reported that when they proudly walked down their street pushing a new baby carriage, neighbors approached with fear, expecting to see something red and scaly inside. When the neighbors saw a normal baby girl, all fears instantly dissipated. "It's just helping Nature along a bit," father John Brown said.

Since then, more than forty thousand American babies have been "helped along a bit" through various forms of assisted reproduction, and perhaps fifty thousand in North America,

---

Originally published nationwide in Knight-Ridder newspapers such as the *Atlanta Journal-Constitution*, 25 July 1998, 18.

Australia, and England combined. The Vatican perversely persists in condemning this procedure, but it is hard to see the wisdom—and certainly not the compassion—behind its thinking. Perhaps no children in the history of the human planet have been more wanted and loved. And there have been no higher rates of emotional insecurity, abandonment, or child abuse among these children. [In 2002, researchers reported preliminary findings that IVF children suffered twice the normal rates of heart and kidney abnormalities and cerebral palsy. They noted that 91 percent of such babies were still normal; to infertile parents, conception of a baby with a 91.4 percent chance of being normal versus 95.8 percent for normal conception still looks good. Another study of children conceived through IVF and aged eleven to twelve found them to be emotionally mature and normal.]

But the story is not altogether as happy as the for-profit fertility clinics advertise on their web pages. Few states require insurance companies to cover in vitro, and the costs run around $8,000 per attempt. Most couples try two or three times but end in failure. In fact, only about 15 percent of couples take home a baby through such procedures, even after paying $25,000 for three attempts.

So the field has looked for new ways to help the infertile. The hottest thing today is finding young women who will consent to having their eggs removed to help an older couple conceive. Such eggs, mixed with sperm from the older man, create an embryo that can be gestated by the older woman. The resulting child at least has a genetic connection to the father. In some clinics, this procedure has doubled and tripled rates of success.

But there is a disturbing, albeit predictable, tendency to report each new possible advance in assisted reproduction in breathless, the-sky-is-falling prose. Whether it is with sperm donations from geniuses, egg donation, babies from thawed eggs, babies from frozen embryos, or a delayed twin from a frozen, previously twinned embryo, each new turn in this reproductive road is reported as if the dreaded slippery slope is just ahead, waiting to take us to the pit below.

If there is anything that we should have learned in twenty years, it is that parents want the best for their kids and don't want to spend years trying to conceive in expensive ways to create damaged children. Nor do scientists and physicians in reproductive medicine.

Lately, the world's fears and imagination have been captured by the possibility of cloning a human. This has been one of the worst cases ever of mindless, thought-stopping reporting. Indeed, the word "clone" is so inextricably associated with being a copy and indeed, with multiple copies, that the word should be banned. A child originated with the same genome as a human ancestor will share the same genetic foundation with the ancestor, nothing more, and will be a delayed twin, nothing more.

Looking back, everyone missed the real issue, which concerned money and the potential it causes for deception, fraud, and greed. Real abuses occurred when Cecil Jacobsen in suburban Washington, D.C., mixed his own sperm with patients' eggs to create many embryos, and when Richardo Asch in southern California implanted patients' embryos without their consent in other, unsuspecting women. The field has also suffered for decades from deceptive statistics and advertising about rates of success in inefficient clinics.

What we can predict for the future is that the money issues will be more important than the emotional ones. A child will be originated somewhere by somatic cell nuclear transfer (the technical name for "cloning") (India may next try this to prove it deserves a place at the world's table). Such a child will likely be normal and soon the novelty will wear off. Not many people will ever use this procedure because creating babies in the old-fashioned way is too much fun and because somatic cell nuclear transfer can only be done through in vitro fertilization, which, as said, is expensive and inefficient.

When somatic cell nuclear transfer succeeds, the money questions will arise. Will any insurance company pay for the procedure?

What if a family has a history of terrible genetic diseases—wouldn't it be much cheaper for an insurance company to pay for a couple to give a child a good genetic foundation than to pay for lifelong care for an afflicted child?

More important, will society allow people to sell their genomes to other couples desiring the same genetic foundation? If we allow payment for sperm, eggs, and embryos, can we deny it for genomes? Shouldn't the owner of the genome make something in the process?

Good answers to such questions are difficult to come by, but if the history of bioethics in creating babies has taught us anything, it is that such questions about money will be more substantial than breathless predictions about dangers to the family.

10

The McCaughey Septuplets: God's Will or Human Choice?

Americans and the national media are rejoicing that all seven of the McCaugheys' babies were not only born, but born healthy. Yet for all the coverage that this story has seen, rarely does the darker side emerge. Call me a curmudgeon but I think something is not right here.

We humans have a dangerous tendency to overgeneralize from one well-publicized case (witness the ban on commercial surrogacy after the Baby M case). In France—where pregnancy is sometimes pursued with an almost religious zeal because each new (French) baby means a bonus from the government—use of fertility drugs has increased the number of triplets tenfold since 1982 and the number of quadruplets thirtyfold.

Some have complained about the costs to society with so many babies in one birth, and it is true that the gestation and birth of the septuplets will probably cost a cool million dollars. Others have complained that the human uterus did not emerge in evolution to bear litters and that large multiple births are unnatural. Still

---

Originally published in the *Birmingham News*, 1 December 1997, C1.

others wonder what toll this extraordinary gestation will take on Bobby McCaughey's body and eventual health.

These are important matters, but they strike me as morally secondary. Costs can be absorbed by spreading them over millions of payers, and what is unnatural in one era becomes normal in the next (witness anesthesia). Also, if Mrs. McCaughey made an informed choice, she is free to risk injuring her body in childbirth as she sees fit.

Still others wonder if the two parents will be able to give each child the nurture and one-on-one parenting that are ideal. Would you want to grow up with one-eighth of the attention you got from your dad? I wonder if the McCaugheys will have the time, energy, and money to allow each child to develop to his full potential.

My concern is about what is really best for the children. This couple took the fertility drug Pergonal, conceived seven embryos, refused to reduce (abort) any, and then said that any results were "God's will." In doing so, they risked the lives and health of some of their babies. They took bad odds and hoped that all seven would be healthy, and in so doing, they took the risk of having seven disabled or dead babies.

Multiple-birth babies are usually premature (each may weigh less than two pounds), are three times as likely as single babies to be severely handicapped at birth, and may have to spend many months in neonatal intensive care units. In a multiple pregnancy, nutrients and oxygenated blood in the womb become a scarce resource (a uterine lifeboat, if you will), where not all of seven fetuses are likely to emerge healthy. To prevent disabilities resulting from deprivation in utero, physicians recommend "selective reduction" of all but one or two embryos.

It seems to me irresponsible to say, as the McCaugheys did, that it would be God's will if any turned out blind, crippled, or dead. If God was clear about anything in this case, it was that the McCaugheys should not have kids, or they would not have needed fertility drugs in order to conceive.

If you take a fertility drug, and if too many embryos conceive, you should be willing to reduce the embryos for the good of the children born. You shouldn't run the risk of severely disabled kids and say it's "God's will" if it happens rather than say it's a grave risk you decided to take.

NBC News recently featured [in 1997] the quintuplets of Denise Amen and her husband, who were offered the chance to reduce and did not. One of her babies was born blind and others are "developmentally slow."

In 1985, a Mormon couple, Patti and Sam Frustaci, conceived septuplets. Informed of the risks of disability and urged to reduce, they refused. Four of their seven babies died, and the three survivors had severe disabilities, including cerebral palsy. The Frustacis then sued their physicians.

And in a 1996 gee-whiz case in England, Mandy Allwood conceived seven or eight embryos at once. Offered a large cash bonus by a tabloid for exclusive rights if all made it to term, Mandy announced she would not reduce any and go for maximal births. As a result, she lost all of them.

I recently heard about a case of a multiple pregnancy in West Virginia where the woman refused selective reduction. As a result of taking such a risk, only one child survived and this child was blind, paraplegic, and severely retarded. The physician on the case said that henceforth he would no longer accept women who would not agree to selective reduction as an option. His position was that (1) he did not get into assisted reproduction to be a physician who helped create severely damaged babies, and (2) although women have the right to be against abortion, they would have to find other physicians who agreed with them and who could accept such terrible outcomes. The University of Nebraska Medical Center is also now encouraging selective reduction in such cases.

Not too many people are interested in long-term follow-up, yet the details that emerge are not encouraging. One physician emphasized on national television that cerebral palsy may not show up until the second year, so it is premature to call all seven McCaughey babies completely healthy. [On their fourth birthday in 2001, the McCaughey septuplets lagged in development and were not all potty trained.[1] Joel suffered seizures; Nathan has spastic diplegia, a form of cerebral palsy requiring botox injections (to paralyze spastic muscles) and orthopedic braces. Alexis has hypotonic quadriplegia, a different form of cerebral palsy that results in weak muscles. Alexis also has had trouble walking and learning to talk and has had an indwelling feeding tube since birth. Another child, Natalie, also still has an indwelling feeding tube at age four.]

Of the five Canadian Dionne quintuplets born in 1934, all seemed healthy at birth but only three are alive today (one died at age twenty of an epileptic seizure). Nor did they lead happy lives, because their parents exploited their fame.

New York City's mayor Giuliani was recently on a call-in radio show when an Orthodox Jewish woman with five little babies (three of them identical triplets) said she felt like killing herself because her babies were driving her crazy. Although the mayor quickly got her help (he was running for reelection), what about all the other parents who don't get free diapers and cars? Jacqueline and Linden Thompson, black parents of sextuplets, were living exhausted and on the edge in Washington, D.C., until a radio caller publicized their plight.

What about the older sister of the McCaughey septuplets? Will her role in childhood be only to help her mother raise the famous septs?

We are a long way in this country from a philosophically consistent policy on fetal rights and reproductive responsibility. The Supreme Court of South Carolina ruled in 1997 in the Whitner case that a mother can be prosecuted for using cocaine in her pregnancy because the usage is presumed to harm her fetus. The National Bioethics Advisory Commission said in June 1997 there

[1] "*McCaughey Septuplets Turn Four*," *Dateline NBC*, November 20, 2001, reproduced at www.msnbc.com/news/660542.asp.

should be a federal law against originating a child by cloning because of possible harm to the baby. Yet others such as the McCaugheys take a terrible risk of having disabled children and we make them national heroes because they invoke God's will and beat the odds. Something seems akilter here.

11

Our New Idol, Life

Once upon a time, there was a country named America, which valued life. In the 1980s, leaders of some conservative religions led the fight on all fronts against passive and active euthanasia. The leaders believed that Roswell Gilbert, whose wife of fifty years suffered painfully from late-stage Alzheimer's and osteoporosis before he shot her, deserved to be convicted for murder and imprisoned, as he was in 1985. These same leaders demanded the same punishment for a woman who shot her declining husband, a physician suffering from the last stages of Huntington's disease; she died years later in the Ohio State Penitentiary.

By the year 2000, elderly patients were commonly living into their late nineties, although with diminishing abilities and alertness. The common experience was a life of rich, vital retirement in the sixties and seventies, followed by rapid mental and physical declines in the eighties and nineties. Medicine by then had become remarkably capable of sustaining even the most ravaged bodies and the world had witnessed impressive new kinds of tissue transplants,

---

Originally published as "A New Parable about Life," *Humane Medicine: A Journal of the Art and Science of Medicine* 2, no. 2 (November 1986): 134–35.

ICUs with computer-controlled delivery of both drugs and nutrition through indwelling catheters, and cryopreservation of dead bodies (albeit without consciousness). A great triumph was Lady Belle, an unknown schizophrenic whose body remained in a continuous coma from 1920 until 2001.

Many of the "Old Old" (those past the age of eighty) lamented their existence, laboring with foreign hearts that could no longer function unassisted, and with brains that had lost 50 percent of their tested IQs. Dependent on others for their most basic functions, obligated to be grateful to aging children for support (if they had any still alive), and expected to watch television perpetually as if this constituted life's meaning, the very old increasingly tried suicide. This proved largely futile because the courts consistently declared that assisting in suicide was felonious.

The elderly also discovered that it was not as easy as everyone thought to commit suicide. After the hand shook and the bullet went an inch astray, they merely succeeded in rendering themselves brain damaged, comatose, or blind. Even with the best of plans, some Good Samaritan would often come along and rush the victim to a hospital. Aides in some nursing homes came to resemble strict generals as they walked down halls looking for clever patients who might devise new ways to die.

Typical of such nursing homes was Old Town, a massive, federally run nursing home in Virginia's Shenandoah Valley designed for a hundred thousand "Old Olds" who lacked living relatives. Old Town practiced standard pro-life suicide prevention: no doors could be locked from within, no windows could be opened (to prevent leapers), and no sharp objects could be found in any room.

More than 33 percent of Old Town's over-eighty population attempted suicide at least once (higher figures occurred on cancer wards); because of insurance regulations, such patients were usually sedated. Pain relievers were generously given in Old Town. Although there was not much joy in Old Town, there was also not much pain—at least of the physical kind.

Those who had tried suicide found themselves in modified straitjackets restraining them from behind into their beds. Those who refused to eat had nasogastric tubes inserted, through which they were force-fed liquid nutrients. Movement sensors were installed to prevent suicides; these were connected to monitors at central nursing desks and signaled to staff when patients ceased lying in beds or sitting in chairs. These substantially lowered insurance premiums for nursing homes.

Formerly, Old Town had had a roof garden where residents would sometimes take their meals, but one blind, diabetic former teacher changed all that. Unable to slash her throat with the standard-issue plastic spoon, she grabbed a server's steel knife, slashed her throat and threw herself headfirst over the railing, hurling herself toward the ground sixteen floors below. Unfortunately, her dress caught on a flagpole at the eighth floor, and she was seen struggling to free herself so that she could fall again and die.

The fire department was quickly called, as were television stations; the TV cameras soon taped a fireman atop the ladder, reaching out to save the woman as she desperately fought him off. The evening news showed both the fireman receiving his medal and the woman in the hospital, safely sedated, straitjacketed, and monitored. After her throat was sutured, she was transferred to a geriatric/psychiatric unit.

In all these cases, and according to the law of the land, a physician had to protect life. A series of laws enacted by the Life Party outlawed natural death, living wills, and assigned durable power of attorney to relatives. Assisting in euthanasia was ruled to be assisting in homicide, and preparations for suicide had to be reported. This law was like honor codes in colleges, which make failures to report offenses as serious as the offenses themselves. The Life Party also engaged in nonlegislative strategies, such as picketing pro-euthanasia and pro-abortion physicians at their homes and offices, scaring their patients away, and tying up physicians in court. They ruined several practices this way and caused most physicians to act so as to avoid being targeted.

Years ago one young couple's third child was a microcephalic infant. As required by the law, any case with potential for infanticide had to be reported to the Department of Life. The Secretary of Life had issued strict orders to Life Officers, one of whom quickly began to monitor the couple's baby and its physicians. Although the child would never recognize its parents, talk, think, or fall in love; although similar efforts had allowed such beings to breathe for as long as seventy years; and although this baby would at best have the consciousness of a chicken, the couple never had the choice of letting their baby die. At age eighteen, their diapered "baby" could not speak and often cried unintelligibly. Both parents worked and their eldest daughter, who forsook college, shared caretaking duties with an expensive nurse sitter. The miraculous triumphs of fetal surgery, fetal monitoring, and intensive care for premature babies led the United States Supreme Court in Reagan's later years to reinterpret its *Roe v. Wade* decision of 1973. Previously the Court had thought (or said it had thought) that viability occurred around twenty-six weeks, but the Court's new interpretation was that in an intensive care unit, viability occurred at twenty weeks. Anticipating further advances and to be safe, the Court set the new cutoff at eighteen weeks. This compromise lasted only one year, during which the new Life Party mounted an all-out attack, sensing ultimate victory. After all, it reasoned, if the Court could tacitly admit error once, it could do so again. And so it was that in Reagan's last year as president, the Court declared that any human life was "potentially viable" and that states could prohibit abortions. Paradoxically, the Life Party also supported America's war in Nicaragua, the death penalty, and nuclear weapons in space.

The last Life Party projects, which were erected outside hospitals, were sixty-foot-tall, glowing, gold cylinders. These cylinders symbolized both that medicine could never renounce life and that even the most shattered shred of life was worth preserving.

So there took place that inversion of values described by Nietzsche; so there stood outside hospitals those golden idols now so infamous; so there began the millennium with that new idol, Life.

12

Twinning Embryos Isn't Cloning

American journalism last fall went breathless for a few days when *The New York Times* reported October 24 [1993] on the first page of its widely read Sunday edition that an infertility scientist had "cloned" seventeen human embryos into forty-eight as a way of increasing the supply of embryos in fertility clinics.

The next day, several other newspapers, CNN, and *Good Morning America* took the story at face value, as did several bioethicists who instantly condemned the event. The *Times* also did two similar follow-up stories. Jeremy Rifkin, who wrote several profitable books saying that the sky was falling because of advances in assisted reproduction, was quoted in the *Times* story as immediately criticizing the event as "opening the door to the Brave New Worlds . . . of human eugenics."

To its credit, *Newsweek* got the story correct in its November 8 issue:

> It is one of the most sought-after coups of 20th century journalism, along with the identity of Deep Throat and

---

Originally published as "Dealing with More Clone Hype," Sunday *Birmingham News*, 20 February 1994, C1.

> Senator Packwood's diaries—the first story that can plausibly use "human" and "clone" in the same headline. . . . Last week, *The New York Times,* based on an apparent misunderstanding of a paper reporting a technical advance in embryology, touched off an echo of the same hysteria with a page-one story whose headline suggested that human embryos were being cloned in a laboratory. Within days medical ethicists were gravely measuring the slipperiness of the slope on which humanity now teetered, while demonstrators marched outside laboratories insisting that no one would ever clone their DNA.

What most people understood by "cloning" was what Woody Allen portrayed in *Sleeper,* the ability to reproduce an identical physical copy of a human being from some cells of his adult body (in *Sleeper,* the dictator's nose) or from dead cells (the dinosaur blood inside a preserved insect in *Jurassic Park*). The *Times* article reported the copying of the undifferentiated cells of embryos by a scientist, Dr. Jerry L. Hall, who used a technique common in stimulating embryos in animal husbandry and easily applied to human embryos. What was not emphasized was that once human cells specialize into skin cells, blood, and so on they cannot be returned to their undifferentiated state to create a copy of the whole body of their donor (from the cells of the dictator's nose, you get, at best, a nose!). After the *Newsweek* article, the story disappeared from the media. *Newsweek* noted that the *Times* "did not respond to numerous requests for comment last week."

We must all be careful not to repeat the folly surrounding the birth of Louise Brown in 1978, when even respected scientists such as James Watson ran together in vitro fertilization with cloning and these in turn with dictator-coerced genetic engineering.

Other breathless, puffed-up "Jurassic Park" stories by the media made people think that Louise Brown would be a monster and almost prevented the development of new techniques of assisted reproduction. This is the stuff of supermarket tabloids. *Newsweek* itself in 1978 hyped the story and made mistakes about the original physician's origins and his carrying around human embryos inside a rabbit. It is good to see that *Newsweek* has now made a 180-degree turnaround.

Between 1978 and 1990, at least 24,000 American babies were born through assisted reproduction, and for every baby born in America, probably two were born worldwide. New methods—such as GIFT (Gamete Intra-Fallopian Transfer) and use of frozen embryos or donor eggs—helped create these numbers. These statistics are the backbone of the successful new field of infertility medicine.

Science editor Nicholas Wade defended the *Times* and argued that Dr. Hall accurately used "cloning" in the title of his paper. Nevertheless, it is not clear that the *Times* did use the term appropriately, because one cannot ignore the cultural context of a story with such allegedly vast ethical implications.

For the millions of viewers of *Jurassic Park*, "cloning" implies creation of an identical, adult organism from the DNA of specialized, developed cells of a living (or dead) adult organism.

This, of course, is not what Dr. Hall did. He repeatedly twinned embryos from undifferentiated cells of other embryos, a relatively easy thing to do, one commonly done with nonhuman animal embryos. Once the embryo's cells start to differentiate, the process cannot be reversed and hence, it is not possible to create, say, another Hitler or Lincoln if we had preserved tissue from their bodies. Hence, many of the ethical issues recently discussed about cloning are nonissues.

Criticizing other responsible news organizations for mistakes is never easy and reporters get few kudos for exposing nonstories. The same is true in bioethics. When reporters called me about this story, I said it was a nonstory and received the cold shoulder. Nobody likes to be told their story isn't a story.

Ideas have consequences, and so do stories and their placement. The history of sensationalistic reporting about assisted reproduction is one of the ignoble stories of the past twenty years of journalism and such sensationalism has contributed to the fears of many Americans about research in fertility and genetics. Politicians remain skeptical of removing the unfortunate fifteen-year-old ban on federally funded fertility research on embryos.

For the 85 percent of the millions of infertile couples for whom in vitro fertilization is either unsuccessful or too expensive, and for those who have a baby afflicted with genetic disease, this ban is a continuing human tragedy.

13

Cloning Michael J. Fox's Embryos

Millions of Americans recently watched Michael J. Fox's last episode of *Spin City* [May 27, 2000], and still others saw the news coverage of a dinner in Washington, D.C., to benefit Fox's new foundation for research on Parkinson's disease. Fox's actions show courage and compassion, and it will be great if he raises plenty of money for this research.

But there is something the federal government can do now to help Fox and others with Parkinson's: allow therapeutic embryo cloning.

Last week, in a lecture to the Royal Society of Medicine, Dolly's cloner Ian Wilmut argued that therapeutic embryo cloning "could be helpful to treat conditions associated with damage to cells which don't repair themselves; there isn't currently an effective treatment for any of them."

Parkinson's is a condition that falls in that category; so are Lou Gehrig's disease (amyotrophic lateral sclerosis) and Alzheimer's disease.

Originally published in the *Birmingham News,* "Cloning to Cure," 4 June 2000, C1.

What exactly is therapeutic embryo cloning, and how might it work to help people with Parkinson's? The first process would be exactly like the process used to re-create the genotype (the batch of genes) of a person, say, Michael J. Fox. A differentiated or somatic cell from his body (any cell could potentially be used) would have its nucleus removed and inserted into an enucleated ("nucleus removed") egg of a woman, which would then be fused together to form a new embryo. The embryo-clone of Michael J. Fox would be a perfect match for the real Michael J. Fox, and nothing derived from it would be rejected by the real Michael J. Fox's immune system.

The next step would be to wait until stem cells develop in the early embryo. Such cells can theoretically develop into any kind of cell in the body. Moreover, they can theoretically do so in the lab, not in a human being.

Parkinson's disease is characterized by a lack of dopamine in the brain. It is hoped that by doing research on embryos, scientists can figure out how to make such embryos develop the kind of neurological tissue that might create dopamine and help people with Parkinson's disease. This might be done either by dripping tissue into their brains or by surgical transplants of brain tissue.

What is truly exciting about stem cells, called in 1999 the "breakthrough of the year" by *Science* magazine, is their ability to repair damaged tissue. This makes their use in research crucial for virtually every branch of medicine.

However stem cells might be used, our current ban of federal funding of research involving human embryos isn't going to help people like Michael J. Fox. Similarly, if we think the sky is going to fall every time cloning is mentioned, we aren't going to get anywhere.

I explained above the reasons we can't just use spare or discarded embryos left over from in vitro fertilization: The embryo must be created to match Fox's body; not just any embryo will do.

Also, the General Counsel for the Department of Health and Human Services ruled last year that government funds can be used for research on stem cells once such cells are extracted from embryos. That is not going to be good enough because there is too intimate a connection between such dot-sized embryos and stem cells; human embryos will need to be used for the best results.

One way of looking at this subject is to ask who the embryos that are derived from Fox's body belong to. Since the embryos should only be taken with his permission and since they are almost identical to his genome, who else could they belong to but Michael J. Fox?

Indeed, I think we should conceptualize research with such embryos not as research on tiny persons, but as growing fresh parts of your own body to heal its diseased parts. It's Michael J. Fox's body; it's my body. Let us do what we want with our bodies and our embryos to heal our own diseases.

Fox's departure from television and his efforts to fight this disease are important because they remind us of what is at stake here. It's not that there aren't some ethical concerns about use of such embryos: conceptionists believe personhood begins at conception, and that it's murder to kill embryonic persons. Others worry about slippery slopes in medical research, where embryonic research might lead to fetal research and then to research on never-competent persons.

These are genuine ethical issues, but they are not the end of the matter ethically. Reasoning also has a role. The question is not whether there are any dangers or abuse or offense to some people but, instead, whether such issues are so important that millions of Americans with diseases such as Parkinson's should be denied potential cures. Is the risk of a slippery slope so great that it should hold hostage such an important area of medical research?

One caveat: At present, research on human embryos is highly tied to abortion politics. It might be such a powerful topic that the only way to lift the ban on federal funding would be at the price of accepting regulation of both federally funded and private research. At present, private research is unregulated and relatively free from abortion politics. Regulation of private research on embryos would be too high a price to pay, because then anti-abortion politics would affect private research. The only good solution is to leave private research unregulated.

Last month, in a hearing on this topic before Senator Arlen Specter's committee, Harvard University's Gerald Fishbach called government-funded research through the National Institutes of Health "the world's treasure." He's right. We need to use our treasure, not put it in a museum.

Last July, a panel of English scientists endorsed going ahead with therapeutic embryo cloning for British patients. Sir Liam Donaldson is heading the English governmental committee that will decide whether such cloning will be allowed, and his committee is expected to decide favorably. It is time that Americans did the same, not only for Michael J. Fox, but for millions like him.

14

A Sheep Is Cloned, Tah Dah!

The word "cloning" is a thought-stopper. Actually, even that statement is premature in that it implies that some thought was begun that has stopped. In the days since the announcement of the cloning of an adult sheep in Scotland, I have been struck at how little thought has gone into discussions of the possibility of cloning a human embryo.

The universal reaction has been Thou Shalt Not! Popular bioethicists measure how far down the slippery slope this event has brought us, and theologians warn of Playing God. People inveigh against medical technology as if the antibiotic they are now taking is not covered by this phrase.

Can we at least talk about the ethics of human cloning?

I doubt it. If history runs true to course, America will immediately ban human cloning, and that ban may then never be lifted.

The same thing happened nearly twenty years ago with the birth of Louise Brown by in vitro fertilization. Jeremiahs said the sky was falling (including Nobel Prize–winning scientists James

---

Originally published as "At Least Can We Talk before Banning Human Cloning?" *Birmingham News*, 2 March 1997, CI.

Watson and Max Perutz). To dabble in assisted human reproduction, most implied, was dangerous, unethical, and against God's will. Eight-cell human embryos were tiny people and should not be used in experiments. Scaremongers such as Jeremy Rifkin created an atmosphere that implied: "Don't trust the physicians and scientists! They are evil and want to create Frankensteinian monsters."

Yeah, right.

But as a result, Congress passed a ban on the use of federal funds for experimenting on human embryos, and that ban still stands. What good did that ban create? None at all. Although a little such research can be done with private funds, 90 percent of all research is federally funded, so other countries have leapt ahead of us in research to help infertile couples.

It is amazing what people say about human cloning. People talk about a person cloning himself. Wait a minute. Feminist bioethicists are certainly correct to emphasize that women tend to drop out of these discussions or just be seen as fetal containers. An embryo can be cloned, but some human female has to gestate for nine months and nurture that embryo for it to become a fetus, and then birth it for it to become a baby. Once a baby, that being becomes a person with a right to life under our laws and does not belong to anybody else. (Doesn't the original source of the DNA own the identical clone? Not at all. No more than the firstborn identical twin owns the DNA of the second. A person owns his own cloned embryonic tissue, but a new baby "owns" himself.)

But couldn't an unscrupulous woman clone and gestate a spare parts version of herself for future use if her kidneys failed? But it wouldn't really be herself.

Take my own case. If I could persuade my wife to gestate an embryo with my DNA in one of her denucleated eggs, the baby that would be born (call him Fred) would have a far different life than mine.

Fred would not grow up in the 1950s in the suburbs of Washington, D.C., but in 1998 in Alabama. Where I knew the Mickey

Mouse Club and 5-cent Cokes, Fred would grow up knowing how to use a computer, watch MTV, and, perhaps most important, would not be the oldest child of five, charged with baby-sitting through much of his adolescent years. Fred's personality would be influenced greatly by my parenting (which is different from my father's) and the parenting of my wife. If I had been looking to clone myself, I would be disappointed.

The movie *Multiplicity* with Michael Keaton contributed to false views when several copies of Keaton's character appeared overnight, without any human gestation in a womb or any development over time. Nice magic.

In my book of ethics, whether or not human cloning would be a good or bad thing depends on two things: On the motives of those who want to do the cloning and on the consequences. It would be silly to prohibit all such acts in advance before we know what we are prohibiting. Yet a bill banning human cloning already has been introduced in at least one state legislature.

Consider a couple where the woman is at risk for a genetic disease such as Huntington's. The couple employs assisted reproduction, creates five embryos that are genetically different, tests them for Huntington's, then implants the three embryos without Huntington's. One embryo is successful and they have one beautiful, brilliant, wonderful child who is the joy of their life. Tragically, when the child is five years old she and her father are killed in a car crash.

The mother has cord blood from the birth of her child. She still ovulates and one of her eggs can be used to grow the DNA of the deceased girl. Are we certain that this would be a morally wrong act?

Someone might object: She hasn't accepted the death of her child and mourned her properly. She's wrongly trying to replace it. OK, that's a bad motive and we can talk about that.

But suppose her motive is not bad. Suppose she says, "Look, I know it wouldn't be the same child and I would never give it the same name. But why shouldn't I create a child that represents my love for my husband and my daughter? With both their genes? I don't want to remarry and I don't want the child of any other man. Why is it better for me to be without children?"

I'm sure there might be objections to this scenario, but at least the discussion would then be of a real woman with real motives, not some vague, unreal, unfactual fantasy.

At bottom, most of the billions of humans who have lived on our planet have had a conception that was random, unchosen, without reflection, and mindless. We pay so much attention to a medical advance that gives humans more choice, while consistently ignoring the misery of the status quo. For example, over the past two decades, a million American teenage girls have become pregnant every year (in which most of the conceptions were random, unchosen, without reflection, and mindless).

Our first, emotional, unthinking reaction is to adopt the fatalistic view that we must accept whatever comes along from the genetic roulette wheel. This view is associated with a conservative religious view that our human nature is flawed, that we are sinners, that we should not tamper with nature, that we should not rise above our lowly place, and that to make any change may place us on the dreaded slippery slope.

In contrast to that immediate, emotional response is the eighteenth-century Enlightenment view that elevates human reason and its capabilities. Also in contrast is American pragmatism, which is skeptical of discussing ethics too far away from the needs of real humans in real cases. And also in contrast is all of medicine, which rejects the fatalistic view of congenital disease as God's will to be endured and the doomsaying view that we do not have the ability to use medical advances wisely.

15

Ban Sexual Reproduction!

"Our twenty-second century should not begin," Senator Greenback declaims, "by abandoning our traditional and safest way of creating babies through cloning." So he opposes adding the antidote to America's water that would revert us to sexual reproduction.

He's correct and here's why.

Human sexual reproduction (HSR) produces unwanted children. Unless we take great care in having sex, children result. Neither planned nor wanted, such children burden families. In contrast, asexual reproduction tests desires of people to have children. Cloning still requires in vitro fertilization, with successes of only 30 percent, so many would-be parents fall by the wayside. Only the luckiest and most persevering get children, as it should be.

Of course, such lucky parents love their children. They neither take for granted their children's existence nor regard them as purchased commodities. Today's cloned children know they were enthusiastically desired, whereas children created sexually always lacked proof of being wanted.

Originally published as "Future Shifts Reproduction to New Level," Sunday *Philadelphia Inquirer*, 2 December 2001.

HSR children will not be fully persons or fully ensouled. The highest levels of human functioning and ensoulment come from expression of the best genotypes. Anything less is subhuman, inferior, and primitive. We shouldn't descend again to the animal level of sexual creation: that is unimaginable.

HSR is against God's will. Why did God allow us to create children asexually if it wasn't part of his divine plan? To revert to sexual origination is to throw a Gift back in the face of the Giver.

HSR oppresses women. If children can be created sexually, men will constantly demand more sex. Also, divorcing sex from reproduction made it fun, not reproductive work, as in the past. Finally, millions of women in the past became involuntarily pregnant through rape, contraceptive failure, or bad timing.

HSR children will be lobotomized for organ harvesting. Because of the above reasons, HSR children will be seen as inferior (as they will be), and hence, will be harvested for organs for their relatives. They will also be slaves in off-world mines.

Sexual reproduction harms babies. Before human reproduction came under the caring protection of physicians, almost any young woman could have children. Many drank alcohol and smoked during pregnancy, causing cleft palates, fetal-maternal alcohol syndrome, and low birth weight, I.Q., and APGAR scores. Thousands of babies in North America were harmed this way.

HSR children may have deleterious genes. With no clinic for assisted reproduction, no social checkpoints prevent adults with Huntington's, cancer, and hemophilia from creating children. In the twentieth century, 2 percent of children suffered from conditions linked to bad genes, fully 120 million people. In contrast, asexual reproduction eliminates the random mixing of genes. Do we want to return to the atavistic state where the health of a child depends on the spin of the genetic roulette wheel?

Sexual reproduction allows narcissists to have children. When asked why they wanted children, HSR parents replied: "To have something of myself continue after death. To have someone inherit my name. To have someone to take care of me in my old age." Procreation from such self-centered motives should obviously be a federal crime.

Society will be prejudiced against HSR children. By now, cloning is so accepted that a child created in a different way will suffer discrimination at school and work. He will be marked as unwanted, as composed of random genes, and as inferior.

HSR children will be raised in batches to be assassins. Without the watchful eye of physicians in a clinic, unscrupulous parents will create children for evil ends. Because drugs creating superovulation are easily obtained, introducing sperm can create seven babies (as in the McCaughey case). Such batch babies raised in the same household will be brainwashed into being assassins or, even, vegetarians.

Sexual reproduction hurls us down the slippery slope. If we allow people to have children sexually again, it will be very difficult to prevent the next step, which will be allowing people to choose how to originate children and when. Next, people will want to choose how many children to have. After that, parents will demand the rights to decide whether to have any children at all.

Unsavory scientists and physicians right now are waiting to help parents down this treacherous path. We must resist going down one step further, lest we soon find ourselves in the pit below.

16

Please Don't Criminalize Human Cloning

When we watch *Snow Falling on Cedars,* we like to think that we would have defended persecuted Japanese-Americans. Had we lived in Alabama in Martin Luther King's time, we like to think we would have marched with him to end racial segregation. Fighting prejudice is easy over a battle long won.

It is less easy to fight prejudice in the present or even to see it in one's world. Yet prejudice surrounds discussions of originating a person by cloning, and few people have risen to fight it.

Consider the many commentators who talk about whether "the clone" would be mistreated. "Clone" is so pejorative that its very use indicates question-begging attitudes, like a report about equal rights for women that begins, "The chicks say they want. . . ."

We are in a similar situation now about cloning humans. Much prejudice and ignorance abound, abetted by movies such as *Multiplicity, Blade Runner,* and in an even worse case, the movie starring Arnold Schwarzenegger, *The 6th Day* (Arnold goes away from home

---

Originally published as "If Cloning Becomes a Reality, Should It Be a Legal Option? Yes." *Insight Magazine,* 25 September 2000, 40–43. Reprinted with permission of *Insight.*

to discover he has been replaced by an identical copy of himself, complete with memories and habits. Even Arnold's wife can't tell them apart. But who is the original and who is the clone? People who don't know much about cloning want to know.)

Most objections to human cloning do not focus on possible physical harm to the new child but instead raise psychological, religious, social, and ethical objections, where prejudice and ignorance are seen.

The *Report on Cloning* by the National Bioethics Advisory Commission (1997) mentioned that a "massive majority" of Americans fear human cloning. Other critics say that "almost everybody" thinks it's a bad idea, and that opinion polls are "nine to one" against cloning.

This is just the ad populum fallacy. Most people wanted to inter Japanese-Americans during World War II, most Americans accepted segregation before the civil rights movement, and too many people today still think that it's bad to be gay or lesbian. Most people twenty years ago feared test tube babies and genetic therapy before these were understood.

A common objection is that originating a person by cloning will undermine human dignity because the new child will not be unique but a copy of an ancestor. This objection is wrong on three counts. First, the mitochondria of the egg in which the ancestor's nucleus is housed will contribute about 1 percent of the genes, so the new child will only be 99 percent genetically the same.

Second, only the ancestor's genes get copied, not the person who was formed with memories, character, free will, and a family. Finally, the objection assumes that the only way a person can be morally valuable is if her genome is unique. So identical twins and triplets lack moral value?

Moral value comes not from a random spin of the genetic roulette wheel but from how society creates moral rules about how to treat humans. Such moral rules must be grounded on morally relevant characteristics of beings, such as feeling pain or consciousness, not skin color, sexual orientation, or method of origination.

"But the child will be harmed because it will know that it was not wanted for itself but for the characteristics of the ancestor." So goes another common objection. According to the Alan Guttmacher Institute, two thirds of American women will be unintentionally pregnant at some time during their lives and many of them will not get an abortion. So already many American babies are unplanned, not wanted for themselves.

A child created by cloning would not need to know how or why it was created. Critics assume it would know, but critics in the past assumed similar things about test tube babies—assuming, of course, that it was a terrible thing to reveal to an adult the emotional and physical ordeals and financial sacrifices that the infertile couple went through to create the child.

But let's bite the bullet and suppose the cloned adolescent is told the truth, that is, he's told that a certain genotype seemed desirable to the parents. "Your Uncle Frank was a great guy and never had a sick day in his life. He loved skiing, played a great saxophone, and graduated at the top of his class at West Point." Would knowing that kind of information harm a teenager?

Before you answer, let me tell you how millions of children were created in most of history. My own grandmother, Vernie Burner, was one of ten children born to a Mennonite Shenandoah Valley farmer who needed a large family to help him cut hay, pick apples, and feed the chickens. Back then, children were created to be farm hands. Parents didn't save for their college education, take them to gymnastic classes, or spend every night at soccer games.

Ironically, when social critics of today look back to our American past, it is precisely this era they see through rose-colored glasses as a time of great American character, true moral values, and family strength. Regardless of whether they are right, most children back then were not created because they were wanted for themselves.

Another example may put cloning in perspective. Modern societies could do something right now that would prevent great harm to millions of children. Evidence for this harm is incontestable and admitted by virtually everyone. This is the harm of allowing women to smoke and drink while pregnant. Incidence of cleft palate was recently shown to vary directly with number of packs smoked per day by the mother. If we chose, we could make it illegal for a woman to do so as soon as she learned she was pregnant.

But we won't, and not just because of the American Civil Liberties Union. We wrongly think such harms to children are normal and that nothing really can be done to stop them. Instead, we focus our fears, novels, and science fiction movies on the sensationalized harm of cloning.

The geneticist J. B. S. Haldane once quipped that in such debates, "An ounce of arithmetic is worth a ton of verbal argument." He was right. In fact, originating humans by cloning requires in vitro fertilization, which is expensive (about $8,000 per attempt) and inefficient (only twenty in one hundred couples ever take home a baby created this way). In America during twenty years, 50,000 babies have been created through assisted reproduction. If the same figure holds for the future, then when cloning becomes safe, maybe five hundred babies will be originated by asexual reproduction of the genome of some ancestor. And all embryos created this way need to be gestated for nine months by a real woman, who might miscarry or change her mind and abort.

So a small number of people will be created by cloning and when they are, we will have evidence to think we can do so without physically harming such children. Rather than convene national commissions to inveigh against such imaginary evils, we might do better to form them to combat the real ways children are born harmed every day.

Constitutionally, there is no basis for the National Bioethics Advisory Commission or the federal government to tell married couples when and how to have children. Nothing gives the U.S. government this power in the U.S. Constitution or Bill of Rights.

The only possible basis is the Interstate Commerce Act. Even if sperm and eggs occasionally cross state lines while being shipped to infertility clinics, it is preposterous to think that such conditions were what Congress intended in creating such an act, much less that such an interpretation would pass review by the U.S. Supreme Court.

But is there any good reason a couple would want to originate a child by cloning? First, notice how hypocritical we are in putting cloning to this test and not the teenage couple next door. If each of our parents had to pass a test of good reasons before some governmental committee, most of us would not have been born. Second, what counts as a good reason will vary with all the different conceptions of the good life that are found in philosophy. Many people now have children so that "something of me lives after my death" or so "someone will take care of me when I'm old." These reasons are self-interested.

There are two great, positive arguments for why human cloning should not be legally banned as a medical option. One day soon, after studies on chimpanzees and monkeys prove that cloning will be safe for resulting children, these arguments will become important.

The best argument is to prevent terrible genetic disease. Consider an Orthodox Jewish couple who had three children die of Tay-Sachs disease. They are reluctant to conceive again and to risk another child dying of this disease. Having children is very important to them, but abortion of embryos and using a surrogate mother are not options. Using cloning, the father's genotype could be inserted into the egg from another woman and the wife could gestate that embryo. In this way, a lineage going back a thousand years could be preserved and continued.

Consider also a man who is azoospermatic and his wife, who has no viable eggs after endometriosis. They could create two children, each a copy of one parent's genotype, and have the wife gestate each. In this way, they could have a genetic connection to two children.

In public policy, we think it a good thing that parents are biologically bonded to their children. If so, why isn't cloning permissible?

We should be very careful before we ban a possible medical option. Twenty years ago, Congress put a ban on federal funds used in any research involving human embryos and we now struggle to lift it (and perhaps we will not be successful). A hundred years ago, a brave Alabama physician named J. Marion Sims first inseminated a wife with her own husband's sperm to overcome infertility. Sims produced a pregnancy, but he was universally condemned and had to stop. In the 1930s, Robert Latou Dickenson tried this procedure again but was also forced to stop. Not until the 1960s did society accept this medical option. As a result, thousands of couples were denied children they might have had.

In such a situation, like the early days of AIDS, it is premature to criminalize a possible, future medical option. We study history in order to not repeat our mistakes. Let's hope we study well.

17

Why Science Fiction Distorts Views of Cloning

I believe that most writing in science fiction (SF) about human cloning misrepresents the facts and, hence, perpetuates misconceptions. As one philosophically favoring the right of couples to originate children by cloning, I groan when I hear so many ordinary people repeat falsehoods right out of SF stories.

I think that science fiction should be both well written and scientifically accurate. I also believe, and psychological research has shown, that the greatest creativity occurs within a tight structure. Too much writing in science fiction violates the tight structure of The Facts.

A typical misconception is that cloned humans would be created in identical batches. In Damon Knight's (1964) short story "Mary," a woman cloned to be a weaver must resist dictatorial genetic conformity (the story reads like an early version of the recent movie *GATTACA*). In *Where Late the Sweet Birds Sang* by Kate Wilhelm (1976), cloned humans are raised in batches and have group intu-

---

Originally published in *Future Orbits: The On-Line Magazine of Science Fiction* 1, no. 2 (December 2001/January 2002), 1–2.

itions. Ursula Le Guin's (1969) "Nine Lives" also has cloned humans not only living in standardized batches but also living as adults in batches. As such, they share an eerie group empathy. In "Nine Lives," nine of a batch of ten clones die, leaving one behind, emotionally devastated. Greg Egan's "The Extra" (1990) has batches of clones of the body of a plutocrat created every five years.

The idea of being raised in batches makes many false assumptions, foremost being an artificial womb. Since 1970, I have heard false predictions about such wombs and to date, little progress has been made. Barring artificial wombs, real women must gestate each fetus for nine months, introducing all kinds of variability merely from differing uterine environments. And despite all the hype, if a premature baby hasn't developed lungs, it will die.

Could a couple create a batch of clones? Not likely. If I wanted to re-create the genotype of my funny Uncle Harry, why would my wife want to gestate five or six other babies at the same time? Indeed, we now know that the womb cannot support more than two or three fetuses without creating a likely disability in at least one. Guidelines now call for no more than two embryos to be introduced by in vitro fertilization, which of course is a necessary step in cloning.

Actually, SF is ambivalent about raising humans away from the natural uterus, so another message is that babies raised artificially will be abnormal. In *Alien Resurrection* (1997) starring Sigourney Weaver, freakish versions of Ripley's body grow in clear, see-through vertical vats. (Why are these rejected fetuses being kept around years later? Why are they always grown in clear vats? The answer is found in the movie version (1978) of Robin Cook's *Coma*: for visual impact.) So powerful is this theme that now everyone expects cloned fetuses to be abnormal and ignores the fact that 25 percent of human embryos created sexually are abnormal and, hence, fail to gestate.

A very popular misconception is that re-creating the genotype re-creates the phenotype and even the actual mind of the ancestor.

Cloning merely re-creates the genes of the ancestor, not what he has learned or experienced. (Even at that, only 99 percent of those genes get re-created because 1 percent of such a child's genes would come from those in the egg—mitochondrial DNA.) Conventional wisdom holds that about half of who we are comes from our genes, the other half from the environment. Cloning cannot re-create environment; it also cannot re-create memories.

The false belief that cloning re-creates a person stems in part from the common, current false belief in simplistic, genetic reductionism, that is, that who a person really is is determined only by his genes. No reputable geneticist or psychologist believes this.

The idea that cloning re-creates personhood is one of the most egregious mistakes, but also one of the most popular because it is the most fun. It allows writers to explore various versions of me, such as the ancient good twin/bad twin theme, which John Varley used in his story "The Phantom of Kansas" (1976), in which a woman grows to love her assassin, her jealous clone. Pamela Sargent's "Clone Sister" (1973) has children created in a batch, with the boys finding their ideal sexual mate in their one cloned sister, Kira. The same misconception also allows actors such as Arnold Schwarzenegger to play multiple roles in *The 6th Day* (2000). Greg Egan's "The Extra" flips this theme when a brain transplant goes awry.

To construct a functioning mind, billions upon billions of unique neural connections must occur. The brain is most plastic from birth to age two, when languages can be most easily learned and when stimulation increases intelligence. Later, a functioning adult human brain can no more be copied or transplanted than could the Internet. References to software and blueprints are only skeletal analogies. Too much science fiction is marred by this terrible scientific gaffe.

The movie *Multiplicity* with Michael Keaton contains most misconceptions about cloning, including instant copies of both adult bodies and minds. No gestation or human growth is needed for human cloning, just overnight transformations. Even in Ira Levin's (1976) *The Boys from Brazil,* things get fuzzy when it comes to how the women are forced to be impregnated and gestate little Hitlers.

Of course, cloning a human genotype does not copy a person, because nurture contributes so much to personality, as well as variations from mitochondrial DNA from the egg and random inactivation of genes on the X-chromosome. Ian MacLeod's "Past Magic" (1990) implies that one could clone the genotype of an unwilling father to re-create the father, including his personality and relationship with his daughter.

Over and over again, science fiction tells us that humans created by cloning will be exploited or subjected to prejudice. Such beings have strange names, such as "Azis" in *Cyteen* (1988) or the android "Replicants" in *Blade Runner.* Such popular views infiltrate us all and become part of the knee-jerk reaction to human cloning. No wonder Dolly's cloner Ian Wilmut (incorrectly) said that beings originated by cloning human genotypes would be treated with prejudice by society. President Bush's reactionary bioethicist, Leon Kass, once said, "It is not clear at all to what extent a clone will be a moral agent."

Another old theme is cloning human genotypes to provide organ sources for the ancestor. In the movie *The Resurrection of Zachary Wheeler* (1971), cloned humans are harvested for organs for their ancestor. And, as mentioned earlier, Greg Egan's brilliant short story "The Extra" plays off this theme: here a rich, sordid ancestor has created dozens of subhuman clones for spare parts, but an experimental brain transplant goes awry, giving an "Extra" the better part of his mind and rendering the ancestor a retarded extra. Call this, and the ending of *Where Late the Sweet Birds Sang,* revenge of the clones, which is great, because it counters some of the standard, prejudicial endings.

Why couldn't we grow a copy of our bodies and use it for organ parts? Or lobotomize a fetus at birth, letting its body grow for spare organs? Answer: We can barely do an experiment with federal funds in America on a human embryo, and never on a three-month-old fetus (in enlightened England, embryonic experiments can only occur up to fourteen days of growth). As stated, any child born of humans would be a person. You can't kill babies at birth and turn them into organ sources. This is murder; because of the Americans with Disabilities Act, even babies lacking 99 percent of their brains can't be killed.

Too much SF also presents people who originate babies in new ways as evil. An overworked theme is the evil biotech corporation, for example, Genetico, Incorporated, in Ken Follett's *The Third Twin* (1996). A secret attempt at genetic enhancement of humanity goes wrong when the third person from a twinned embryo turns out to be a psychopathic killer. In C. J. Cherryh's trilogy *Cyteen* (1988), the Reseune Corporation creates cloned humans who travel to planets that it takes decades to reach and who therefore cannot in their lifetimes return to Earth. Azis, programmed in childhood under hypnosis by tapes to do jobs that match their genotypes, are exploited, and Cyteen decries this injustice.

In Michael Crichton's *Jurassic Park* (1993), John Hammond's Dinosaur Park Company overcomes its skepticism about the venture because of its greed. In *When the Wind Blows* (1998) by James Patterson, the biotech/IVF clinic is the evil corporation, complete with a SWAT team that must hunt down the escaped, angelic children with genetically added wings. Take-home message: Evil motives lurk in corporations that would clone humans.

This really hurts real scientists, who are just people. Most of them have kids of their own and care a lot for children. No one wants to bring a handicapped child into the world. Movies and novels rarely portray life scientists with any sympathy. This anti-science prejudice started with Mary Shelley's *Frankenstein* and continues with nefarious scientists working for the government on *The X-Files*.

It is therefore easy to understand why ordinary people are so afraid of cloning. Politicians of course always use corrupt scientists to produce evil results as seen in Ira Levin's classic, *The Boys from Brazil.* Even scientists with good motives underestimate the inherent danger of biotechnology and produce evil results: *Jurassic Park, The Lost World* (1997).

Another perennial theme is that parents who originate human genotypes have selfish motives. In *The Cloning of Joanna May,* by Fay Weldon (1989), a rich man attempts to clone his dead wife.

Occasionally, rays of light penetrate the darkness. In *Cloned Lives* (1976) by Pamela Sargent, a man has good motives in cloning the genotype of his dead daughter. This novel basically is about discrimination against people originated by cloning. But they still are created in batches.

One of the most powerful assumptions of our time is genetic reductionism, especially the simplistic, one trait–one gene interpretation. Running against this produces interesting results. In Charles Sheffield's "Out of Copyright" (1989), genotypes of great geniuses, such as nineteenth-century geneticist J. B. S. Haldane, are cloned and auctioned off every year in an event like the draft of the National Basketball Association. Realistically, the new human version of the genotype often lacks the great characteristics of the ancestor because those characteristics will be such a subtle, fragile result of many neural, uterine, experiential, and historical interactions with the evolving genotype. Even this short story perpetuates the misconception that cloned humans could be bought and sold.

That is because a great assumption of many people is that humans originated by cloning would not be fully ensouled. The same false rap predated babies created by in vitro fertilization; ditto, babies born of mixed race, gay/lesbian, and unmarried parents. Ethically, once a child of human genes arrives and looks, acts, and feels like a person, he is one; legally in most countries, human birth confers rights to citizenship, inheritance, and medical care. Lawyers and politicians opposed to cloning call cloning "replication" or "commodification" because they know that if cloning is considered just another form of human reproduction, it is protected by many U.S. Supreme Court decisions. Only if you assume that a cloned baby is subhuman can you seriously consider using it for organs. Incidentally, *Blade Runner* (1982) flips this mistake: intended to lack souls, replicants develop them anyway and we empathize as they are hunted down and killed.

All in all, I am not proud of science fiction writing about human cloning. For every good piece, there are dozens of unfactual ones. For every good theme, there are all the bad ones. Who is going to write the first great novel about human cloning that does not violate what we know? It remains to be done.

18

If Parents Expect Bad Things from Cloning, Should We Ban It?

The main moral objection to attempts to originate a child by somatic cell nuclear transfer (SCNT) concerns possible harm. Although some critics worry about vague harms such as possible harms to society or to the institution of the family, the main worry concerns harm to the resulting child himself.

This harm is of two possible kinds: *physical* harm as a birth defect or as a latent genetic dysfunction, and *psychological* harm. Psychological harm is often claimed to be caused by parental expectations about the nature of such a child. In particular, there is a widespread belief that any child originated by SCNT would be harmed by unrealistic expectations created by his parents in comparing his future to the life of his genetic ancestor. I call this the argument from parental expectations against SCNT.

One version of this argument claims that what is wrong about SCNT origination is that it is wrong for a parent "to want a copy of himself." I shall call this the wrongness of self-replication objection.

Originally published in "Parental Expectations and Cloning," *Newsletter on Medical Ethics of the American Philosophical Association* 98, no. 1 (Fall 1998).

This popular objection is vulnerable on two points. First, the objection is not about SCNT but about having the wrong motives for creating any child, not just a child created by SCNT. The worry here is about vanity or narcissism, not about SCNT itself. In this worry, in the mind of the objector, the only kind of child who is anticipated to be created would be from the genome of one of the parents. But if this were the real problem, then the objector should have no problem if the parent chose the genome of someone else, say, a favored uncle or a brilliant aunt. But the worry does not go away if this alternative is suggested, allowing us to infer that the objection is not to SCNT itself but to self-centered motives in originating any child. As such, this is a criticism that applies to many cases of human sexual reproduction as well.

It may be helpful at this point to explain two key terms. The *genome* of an individual is the complete set of his genes. How a genome is expressed in a particular individual is his *phenotype,* the result of the interaction of the genome with the environment. An individual's phenotype is the resulting entire physical, biochemical, and physiological makeup of the individual. Exposing the same genome to different environments creates different phenotypes, such as where one fetus receives inadequate nutrition during gestation and another superior nutrition.

To return to the first objection about parental expectations, the other problem is that this "wrongness of self-replication" is based on a falsehood. The idea that the phenotype of a child originated by SCNT would copy the phenotype of the genetic ancestor is false, and false in a myriad of ways. We know that so-called identical twins have tiny differences in their genomes, caused partly, probably, by how much of the X chromosome is inactivated in fetal development, a random process in each twin. Such differences, writ large twenty years later, may account for why even conjoined twins, such as the famous Eng and Chang, may have opposed personalities, to the extent of one being alcoholic and the other a teetotaler. We also know that the gestational mother of an SCNT-originated fetus will contribute a small number of genes to the resulting child, meaning that the final genome of the resulting child will differ slightly from that of the genetic ancestor. Even at its most basic level, the SCNT child's genome will never be an exact copy of its ancestor's.

Beyond the genome, the new child will have a different environment from the ancestor in not just the obvious ways of place and time of upbringing, specific parents, schools, and friends, but also in less obvious ways, such as the gestational mother's diet and use of nutritional supplements and how much the parents talk to the newborn during the first two years of its life (resulting, some think, in how many neural pathways for language are formed). So, in conclusion, the SCNT child would be neither a genetic nor a phenotypic copy of the genetic ancestor.

Nevertheless, some people who want to originate an SCNT child this way will believe they will get a copy of themselves. This raises a new, but much more general, question, "How important should such false beliefs be in making public policy about bioethics?"

This question arose in the 1980s when many people had false beliefs about HIV infection. Some believed that HIV-infected children could infect healthy children at school; some orderlies and hospital aides left food outside the rooms of patients with AIDS, fearing infection if they entered the rooms; other fears concerned contamination in public restrooms, from mosquitoes, from communion cups, and from touching currency and coins.

Looking back, it was wrong to let public policy be made on such false beliefs. When it was, it confirmed the fears and made them seem legitimate. The best way to deal with falsehoods is to act as if they are false while trying to educate people about the truth.

The case is similar with false expectations about SCNT children. Prospective parents will need to be educated about what to realistically expect. In this regard, counselors will play an important role in clinics specializing in this kind of assisted reproduction.

Compare such counselors to genetic counselors. Genetic counselors often see clients at risk for a dominant autosomal genetic disease (such as Huntington's) who have a roughly 50-50 chance of having the gene. When such clients come to be tested, their real motive is often their hope that they will discover that they do *not* have the gene for the disease. Yet the motive of such clients must, unfortunately, be disappointed in approximately half the cases.

An important job of the genetic counselor is to make such at-risk clients understand and feel what they will experience with an unexpected result. When such counseling succeeds, people no longer have a false belief and many don't get tested. Such successful counseling explains recent commentaries expressing surprise that more Americans with such diseases haven't decided to be tested. But why should that be a surprise, given the lack of treatment for these diseases and discrimination by American insurance companies?

Fortunately, SCNT origination cannot be done outside clinics for assisted reproduction. Thus there is a bottleneck and a social check in this particular kind of human reproduction that is not possible for most human sexual reproduction. Thus there is an opportunity to mandate counseling and to make sure that most parents do not have false beliefs about prospective SCNT children. Therefore, because it is generally a bad idea to make public policy based on falsehoods, and because there is a practical way to correct false expectations of prospective parents, this objection to SCNT origination is not a reasonable one.

The objection about the wrongness of self-replication has so far been countered by saying that expectations about self-replication are generally false. But let us now consider another form of this objection about expectations. This one tries to argue not from false expectations but from true ones.

Some people at this point object, "Yes, it is true that the child would not be an *exact* genetic copy, and yes, the expression as an adult would vary a lot, but still, there will be a lot of similarity between the genetic ancestor and the child. And to that extent, the parental expectation will be based on truth. But it would be wrong to create a child this way because, to the extent that he is similar to one who has lived before, the parent will have expectations about him that could, but need not, become true. They could be true if the child has the same upbringing as the ancestor."

This objection against SCNT origination is called the argument for an open future. Because someone with this genome has lived before, and because this child was created in part (or even mainly) because of the characteristics of this particular genome, the future of the first SCNT-originated child is claimed to be closed in a way not true for every other human child that has ever been born. What is wrong is that every child should have a completely open, indeterminate future.

I think this objection can be met in a variety of ways. First, a general comment about the general picture of child-creation that this objection assumes. This vision of child-creation is ultimately Kantian in its assumption that a child should be valued in itself, not as a means to the satisfaction of someone else's desires. As such, a child should be wanted in itself, not for the particular characteristics it might or might not have.

I believe that this sets up a false ideal of not only why people have children but how they ought to have them. I believe it stems first from earlier times in history when people had no control over traits of children and, second, from a very recent trend to put children on an altar to which the lives of parents are often sacrificed, through an excess of developmental activities such as computer camps, special schools, and participation in sports.

I will return to this point later, but what I want to focus on now concerns what exactly is considered to be objectionable about parental expectations about SCNT. Suppose we can assume that parents have been counseled and that, based on primate studies of SCNT, they have good evidence that a child with 99 percent of the genome of Princeton bioethicist Peter Singer or model Elle

MacPherson will be more likely than normal to be intelligent or beautiful. Suppose also that intelligence and beauty are what they value.

One thing that might be considered objectionable is to give parents any choice at all about such characteristics. Many people seem to have such beliefs. The objection here is of two kinds: that such choice is intrinsically wrong and that such choice is indirectly wrong because it will create undesirable consequences.

People who believe that choosing characteristics is intrinsically wrong often believe that it is up to God, nature, or evolution to determine who is born and with what characteristics and that it is wrong for humans to make such choices.

People who make this objection often get confused at this point between objecting that (1) human society should not make choices about which human characteristics are desirable and try to bring about babies with such characteristics, and (2) particular parents should not make choices about which human characteristics are desirable and try to bring about babies with such characteristics. The first is associated with eugenics and dictatorial states taking away reproductive choice from couples; the second is an expansion of current reproductive choice for couples. Fears and concerns about the first are not arguments for curtailing choice in the latter, but the opposite, for a strong and regularly exercised right to reproduce as one chooses is a good check against reproductive coercion.

Most people don't really believe that choosing characteristics in future people is wrong. Parents send their children to one school rather than another based on such beliefs. Prospective parents use genetic tests to test embryos and fetuses to avoid severe genetic disease and abort those testing positive. Mothers avoid cigarettes and alcohol during pregnancy to protect their fetuses.

To argue that such choices are wrong ultimately is to accept reproductive fatalism. Such a fatalism must also apply to who gets pregnant, and it is no surprise that those who oppose genetic choice usually oppose abortion and contraception. So the Vatican opposes all three, consistently.

In the Vatican's consistency, there is an insight. In its view, humans should accept everything that happens as God's will and attempt to change nothing. Each pregnancy is as it should be, and God has a sufficient reason for it. To oppose his will is blasphemy and sin.

The insight is about what happens once human choice is allowed. When Martin Luther wanted to read the Bible and interpret it for himself, the Vatican foresaw what would happen if many Luthers came along, each wanting to interpret the Bible and, ultimately, each wanting his own church. This is also true about human choice. Once humans are allowed to choose anything about human reproduction, there is no logical place to stop.

Once we let people choose to use contraception to prevent pregnancy and to abort unwanted pregnancies, it is hard to justify not letting them abort because a fetus has a genetic disease, because it will be deaf, or because it will be a dwarf. Once we allow abortion to avoid dwarfism or deafness, it is hard to disallow medical treatments at birth designed to overcome dwarfism, such as administering human growth hormone, or deafness, such as cochlear implants.

Implicit in these decisions is the judgment that being short and deaf is undesirable, being tall and able to hear is good. Once those judgments are made, and it becomes possible to choose children who are able to hear better or be taller, it becomes difficult to say why it is wrong to allow parents to make these kinds of choices.

Here it will be objected that some parents will put too much weight on one characteristic, such as intelligence or the current ideal of female beauty, and then be very disappointed when the child does not measure up. We can make this point more generally this way: to emphasize a connection between the value of the child to the parents and the particular characteristics the child may have—characteristics that are uncontrollable, unpredictable, and morally arbitrary—is to harm any child who lacks the desired characteristics or who does not possess them to the degree desired by the parents.

If a parent has such an obsession about a single characteristic, and believes that an SCNT child will certainly have such a characteristic, then what we have here is a variation on the reply to the first general objection (that the expectation will not be realized) because it is very likely false that the child will have a particular characteristic to the degree of the famous ancestor. Moreover, it will often be a particular combination of traits that is desirable, not just a trait such as physique or skill in solving mathematical puzzles, and such a combination will very likely be the result of the child's environment and education.

A different variation here concerns parents who focus on looks in originating an SCNT child. In the famous Baby M case of commercial surrogacy in the United States, the Sterns chose the surrogate mother, Mary Beth Whitehead, only because she had a physique that *looked* like Mrs. Stern's. In this, we all know now, they were very ill-informed, for they disregarded the importance of personality, intelligence, and, perhaps, mental illness in making their selection. Similarly, couples seeking embryos or eggs for implantation in older infertile women in America often choose mainly on the basis of the *looks* of the genetic ancestors.

As the Baby M case taught us, and as other cases of embryo and egg donation are teaching us, to overemphasize appearance to the exclusion of all other characteristics, including personality, intelligence, and overall health, is very dangerous.

Overall, the great lesson of modern genetics is that everything is complex. There may be no single gene for breast cancer but hundreds of variations, so that to tell a woman she has no genetic risk, one must test for each and every variation. Many qualities of the phenotype will be, first, multifactorial at the genomic level and, second, multifactorial at the level of phenotypic expression, such that it will be incredibly difficult to predict or control how any particular quality gets created ("multifactorial" means that three, four, or . . . twelve factors might combine in a unique way to produce a particular trait and that such a particular combination of so many factors will be difficult to re-create). Originating children by cloning will not deprive them of an open future nearly as much as people might think.

But this is just a matter of education and experience. People will need to learn and will need counseling. But I have great faith in prospective parents and what they are capable of learning. The learning curve here will be short: a few well-publicized cases of silly parental expectations will teach thousands, if not millions, about the basic lessons of Genetics 101.

To conclude, to the extent that the argument from parental expectations is based on false beliefs, it can be countered both by saying that we should not make public policy in fear of false beliefs and also that social mechanisms are available to counter the damage of false beliefs of parents regarding SCNT-created children. Even then, some small damage from expectations may occur, but if so, it is likely to be no greater than any other kind of failed expectation of parents about sexually created children.

On the other hand, even if SCNT-created children correctly fulfill parental expectations, such as the SCNT child created from Peter Singer's genome who really does have a talent for moral philosophy or an SCNT child of Elle MacPherson who really looks like her, this does not negate the free will of such children to lead other kinds of lives.

Before we leave this point, consider that people often embrace what they have natural talents for. If you can acquire a great deal of fame and money by doing something that comes naturally, then (all other things being equal) you will probably end up doing that. Most of the people in the world are, statistically at least, average and many desire to be above average in some way. As such, a genomic advantage will likely not be dreaded by the SCNT child in what American bioethicist Arthur Caplan calls the "Oscar Wilde

Syndrome." Instead, an SCNT child may just as likely be grateful for his talent and may even have other SCNT children like himself. In such matters, it is best not to act hastily, as so many did in the past in condemning many forms of assisted reproduction that later proved benign.

If twenty years of experience with such new forms of assisted reproduction have taught us anything, it should be to be open-minded about new techniques, to be skeptical of claims of monstrous babies and of great harm to society, and to think better of parents than to always characterize them in public as selfish and vain. That said, I don't think we have much to worry about from parental expectations about human cloning.

19

Do Not Go Slowly into That Dark Night: Mercy Killing in Holland

The Dutch are known for practicality, fierce independence, moral integrity, free-thought, and defense of civil liberties. Under Nazi occupation, Dutch physicians successfully resisted Hitler's programs. Holland is nominally Catholic, but 27 percent of Dutch citizens claim no religious affiliation and Dutch Catholicism is very democratic and aggressively anti-Vatican. Moreover, the Dutch seem to distrust elaborate technology and wish to die at home more than Americans, 80 percent of whom die in hospitals.

Seventy-five percent of Dutch citizens now endorse mercy killing by physicians after requests by terminal patients. These attitudes have changed Dutch medical practices. It is estimated that in the 1970s and 1980s, Dutch physicians assisted in the deaths of between 5,000 and 8,000 terminal patients. This new development in the history of medicine needs to be studied carefully.

The Dutch national movement for decriminalization began in 1971 when a female physician accepted her mother's repeated

---

Originally published as "Do Not Go Slowly into That Dark Night: Mercy Killing in Holland," *American Journal of Medicine* 84, no. 139 (January 1988): 139–41. Reprinted by permission of Excerpta Medica Inc.

requests for death. Geertruida Postma's mother had had a cerebral hemorrhage that left her partially paralyzed, deaf, and mute. After the stroke, Postma's mother lived in a nursing home, where she was tied to a chair for support. Dr. Postma said, "When I watched my mother, a human wreck, hanging in that chair, I couldn't stand it anymore." So she injected her mother with morphine and killed her. Postma then told the director of the nursing home, who called the police.

Postma was found guilty of murder, but she was given only a symbolic sentence. Postma's lawyer argued that rational suicide assisted by a physician should be a legal defense. The judge rejected this defense but nevertheless specified some guidelines. The 1973 case of physician Piet Schoonheim also concerned mercy killing, but of patients who were not immediately terminal. Schoonheim was eventually cleared.

Two Dutch euthanasia societies grew after such cases. One group sought legislative reform, like the American Society for the Right to Die, and the other helped terminal patients to die, like the American Hemlock Society.

In 1973, the Dutch Medical Association softened its condemnation of mercy killing by physicians. While stating that euthanasia by request should still be a crime, it urged that courts should decide whether physicians could be justified in particular cases. Public support for euthanasia dramatically increased over the next decade, and in 1984 the Dutch Medical Association advocated new guidelines stating that: (1) only a physician may implement requests for euthanasia; (2) requests must be made by competent patients; (3) patients' decisions must be free of doubt, well documented, and repeated; (4) the physician must consult another independent physician; and (5) a determination must be made that no one pressured the patient into his or her decision. Two other guidelines were more vague: (6) the patient must be in unbearable pain and suffering without prospect of change; and (7) no measures can be available that could improve the patient's condition or render his or her suffering bearable.

Dutch medicine had long before rejected the American view that actively killing a terminal patient differs ethically from withdrawing care. In the past decade, Americans have increasingly accepted withdrawing heroic treatment, standard treatment, and most recently, food and water. Groups such as the Hemlock Society now campaign in California for a bill to let the terminally ill request a voluntary physician's assistance in dying.

In 1985, the *Final Report of the Netherlands State Commission on Euthanasia* appeared with thirteen of fifteen commissioners recommending a new exception to the criminal code covering homicide. Mercy killing by physicians was proposed as a legal excuse to homicide, like self-defense or insanity. The remaining two commissioners disagreed because of beliefs about the sanctity of life and the danger of the slippery slope. The final report created emotional debate in Holland. Politically, a majority of the Dutch parliament accepted the recommendations, but the Christian Democratic party strongly opposed them.

Between the spring and summer of 1986, opposition formed to a bill in Parliament accepting the report's recommendations for decriminalization. Opposition was led by anti-abortion groups and Catholic bishops, who called the bill "a descent into barbarity." Evangelical religious groups produced a television show featuring old pictures of victims in concentration camps and emphasizing the slippery slope under Nazi medicine from the initial killing of mental defectives to the Final Solution for *lebensunwertes Leben* (life unworthy of life). They said physicians would push the outer edges of legal euthanasia, generalizing to different but similar cases. They claimed that euthanasia by patient request was murder and that people are the creatures of God who do not have the right to end their own lives.

Champions of the bill retorted that the Nazis practiced murder, which had nothing to do with their cause. They said that using sophisticated medical technology is a physician's right but it imposes an obligation to end life when no purpose is served, that physicians too often merely prolong dying, that the slippery slope metaphor wrongly implies that society cannot make small changes without creating moral nihilism, that physicians are not eager to kill patients, that personal physicians can be trusted, and that if decriminalized euthanasia created bad results, it could be recriminalized.

On February 18, 1987, while the euthanasia bill was being debated in Holland, ABC's *Nightline* staged a live debate between Dutch physician Peter Admiraal; Betty Rollin, author of *Last Wish;* and Father Robert Barry, a religion professor. Admiraal, an anesthesiologist and oncologist who practices in a large Catholic hospital, spoke for Dutch physicians favoring mercy killing. Between 1972 and spring 1987, he had performed mercy killing for more than a hundred patients. Barry criticized Admiraal's work, claiming that if patients were really in "unbearable pain and suffering," they couldn't make rational decisions to die. Barry accused Rollin (who described the death of her mother in *Last Wish*) of a "failure to care" in allowing her mother to die quickly. Barry also accused Rollin's physician of not knowing how to treat cancer pain, and he predicted that if euthanasia were decriminalized, the "immature and the desperate" would increasingly choose suicide.

Admiraal agreed that pain in cancer patients could be stopped and was not the real issue. Instead, Admiraal said (in another interview):

> But there is severe dehydration, uncontrolled itching and fatigue. These patients are completely exhausted. Some of them can't turn around in their beds. They become incontinent. All these factors make a kind of suffering from which they only want to escape.
>
> And of course you are suffering because you have a mind. You are thinking about what is happening to you.

> You have fears and anxiety and sorrow. In the end, it gives a complete loss of human dignity. You cannot stop that feeling with medical treatment. (Alan Parachini, "The Netherlands Debates the Legal Limits of Euthanasia," *Los Angeles Times*, July 5, 1987).

Some Dutch physicians disagreed, including Dr. Karel Gunning, who claims to know "quite a number of cases" where "we thought patients would die in twenty-four hours" but were mistaken and they "lived thirty more years." He said on *Nightline* that a medical prognosis of a terminal condition is only "a guess" about the time of death.

Two more recent developments grew out of the first cases and make some people uneasy. First, the Dutch Medical Association allows terminal children to die even if their parents oppose it. The argument is that dying children do not understand why they have to live to suffer and that prolonging life is too often merely for the parents. Second, Dutch physicians and judges have now accepted other cases for mercy killing covering paraplegia, multiple sclerosis, and gross physical deterioration at advanced age. Some see these developments as a slippery slope, others as logical consistency. In all these cases, patients themselves must repeatedly request a physician's assistance to end their lives.

Contrary to many reports in the media, Holland has not legalized mercy killing. To explain what has happened, I will describe some features of Holland's judicial system that are different from those of the American judicial system.

In Holland, trials are decided by judges without juries; judges are professional and unelected, ruling for life and often making decisions unpopular with vocal minorities. And Dutch district attorneys have a discretionary power to not prosecute that surpasses that of American prosecutors. What has happened is that Holland has a de facto agreement among its powerful judges and public prosecutors to not prosecute physicians who perform (officially illegal) mercy killing done according to precise guidelines.

Performing euthanasia appears to have been accepted as an occasional duty by Dutch physicians. One study discovered that most physicians have two patients every three years who request mercy killing, one of which is performed. Another found that nearly 80 percent of general practitioners in Holland had "experience with" mercy killing. The preferred method appears to be injection of morphine followed after deep unconsciousness with curare.

Famous American decisions about euthanasia differ from these changes in Holland in two ways. First, they concern withdrawing treatment—respirators in *Quinlan,* chemotherapy in *Saikewicz,* and food and water in *Conroy*—not terminating life as such; and second, they concern incompetent patients whose wishes could only be inferred. The American cases should be called "euthanasia by inference of an incompetent patient's wishes," not voluntary euthanasia.

In America, the case relevantly similar to the Dutch cases concerns Elizabeth Bouvia, who had no terminal disease and who wanted to starve to death painlessly with medical assistance. In New Jersey this year, Beverly Requena sued to be allowed to remain to die at a nursing home run by St. Clare/Riverside Hospital. The nursing home had agreed to allow her to be transferred to remove feeding and hydration tubes but would not allow it in their residence. But none of these cases involved active killing, and how Americans feel about such actions is uncertain.

Physicians in Holland dislike the current situation because they are told, "You will probably not go to jail for something which is officially a crime but which you have an obligation to occasionally perform." They would rather have a new statute telling them exactly when they can mercy kill and not be prosecuted. As things stand, misinterpretations are possible. One Dutch physician who failed to document a patient's request for euthanasia received a one-year prison sentence, and three nurses who secretly killed comatose patients received longer sentences.

Moreover, the Dutch program seems to be personally difficult for Dutch physicians. It is rare for them to encounter the idealized terminal patient who is intelligent, in untreatable pain, and with supportive relatives. Instead, they find ordinary people, usually quite alone or with one relative who visits on weekends, who are not in acute pain or at death's door, but who are otherwise very badly off, such as having lost seventy pounds to esophageal cancer. Helping these patients die is not easy. In contrast to a slippery slope of mass killings, few Dutch physicians seem eager to kill.

What will happen in the future is difficult to predict. A statute won't solve all problems because it can't be drafted both broadly enough to cover all patients and precisely enough to close loopholes ("terminal condition," "unbearable suffering"). In any case, the Dutch experience may show whether mercy killing leads to Auschwitz, civilized death, or somewhere in between. [In 2001, after decades of increasing legalization, Queen Beatrice signed a law passed by the Dutch legislature making physician-assisted death totally public and legal. The next year, Belgium followed suit.]

20

Even with a Living Will, It's Tough to Die Well in America

Correan Salter did everything right.

She didn't want to end up in a coma like Nancy Cruzan, an emaciated hostage to medical technology on a feeding tube. So she signed a living will and told her friends that if she ever wound up like Karen Quinlan or Nancy Cruzan, she wanted to die. She didn't get her wish. Given the first court ruling in Alabama about such cases, and one with national significance, almost anyone in this state could end up the way she has.

Eight years ago, the then sixty-seven-year-old married Alabama woman signed her living will. But she lent it to a friend, who's never been able to find it since then. A short time later on August 13, 1987, Correan suffered a massive stroke, destroying much of her brain and leaving her paralyzed and unable to speak. Ever since, she has been in a nursing home in Spanish Fort, on the east side of Mobile Bay. A feeding tube surgically implanted into her stomach keeps her body alive.

Originally published as "Letting Go," *Birmingham News*, 22 January 1995, C1, C4.

At first her husband wanted to give her a chance to recover. He waited two years, but then he died. Then her sisters and brothers became her next of kin.

Last year, the family asked the Skilled Nursing Unit of Westminster Village in Spanish Fort [Alabama] to remove her feeding tube and let her die. Such feeding tubes are removed all the time in Alabama from patients like Salter. Although it is the opinion of the legal counsel of the Alabama Nursing Home Association that current Alabama law permits the removal of feeding tubes, Westminster Village balked in this case. Why is that?

Part of the answer is that the original bill creating a living will does not specifically mention removal of feeding tubes. (And Salter's case is complicated by the missing living will.) However, the 1990 U.S. Supreme Court ruling in the *Cruzan* case said removal of such tubes was not different from removal of respirators.

The cases of Karen Quinlan in 1976 and Nancy Cruzan in 1990 taught us that a person can be dead even if her body still breathes and her eyes are open and can move. The primitive organs and eyes work, but the higher brain is gone, and that is the seat of personhood. Karen Quinlan and Nancy Cruzan were both in a persistent vegetative state, a deep form of coma that is usually irreversible.

One physician testified that it was possible that Salter's body would live over twenty-five years. In the literature of bioethics, the American record in such cases is held by Rita Greene, a nurse who worked in Washington, D.C., and whose body has been in a persistent vegetative state since 1952.

Another reason the nursing home won't remove Salter's feeding tube is that her medical condition is slightly better than that level of coma: her eyes appear to follow some movements in the room. She does not appear to recognize anyone, but her body does respond to very painful stimuli.

Many tests have been attempted to communicate with her, for example, by having her blink her eyes or move them according to some pattern, but all have failed. The best neurologists concluded that she is functionally equivalent to being brain dead. Every medical opinion agrees that her overall condition is irreversible.

Hers is a case that could set a new national precedent—just as the Quinlan and Cruzan cases did. Nevertheless, on December 30, Baldwin County Circuit Court Judge Pamela Baschab ruled that Correan's family could not have the feeding tube removed.

Let's look at two arguments about this ruling. First, at ethics. Salter's case has an either-or premise that supports the moral conclusion that the feeding tube ought to be removed.

On one hand, neurologists testified that she is equivalent to brain dead. In that case, it is senseless for the state to keep her alive. On the other hand, suppose that the neurologists are mistaken and that someone is "at home" in there. That possibility is morally the worse. If Salter perceives existence, like a drowning person looking up from the bottom of a swimming pool, to what kind of horrible torture have we subjected her for the past seven years? What kind of hell does she inhabit? I don't know about most readers, but short of suffocating for seven years, I can't think of any worse way to die. And it could go on for years and years.

Either she's dead and it's senseless to keep her body alive, or she's not dead and it's torture to keep her alive. Either way, it is immoral because it harms her family (by keeping the body alive for no good reason) or it harms what is left of Salter (by torturing her).

This conclusion obviously concerns the lack of any quality of life for Correan Salter—a possibility she wished to avoid. Judge Baschab rejected any decision based on lack of such quality of life, citing "the State's interest in the preservation of human life, an interest that is unequivocal and fundamental." She said, "the State has an obligation to allow only public policies that support the preservation of human life. Any 'right to die' must be viewed in the light of the proper role of government and opportunities for abuse. Once the State accepts a policy that some lives are not worthy to be lived and some people would be better off dead, it takes the first step on a long and slippery slope from which there can be no turning back."

This is an odd argument. Let's pursue its logic. More than two million Americans die each year and about 80 percent at some time will be on a respirator. If quality of life should never be a factor, we should keep all two million dying patients indefinitely on respirators and feeding tubes. A few people will spontaneously recover for a few months and some diagnoses will be discovered to be in error. Although it will cost billions, this public policy is the only one that maximally supports the "preservation of human life." Reductio ad absurdum.

It is the quality of life aspect of this case that gives it potential to set a national precedent. Will the courts ever stop trying to create new ways of saying a person is dead (the easy way out) and start saying a person can die because of horrible quality of life?

Judge Baschab's trite slippery slope argument is always brought out to oppose any change in medicine. It was used to oppose the introduction of anesthesia for women in childbirth at the turn of the century. It was used to predict moral ruin when physicians were allowed to give married couples contraception after *Griswold v. Connecticut* in 1965.

We were also predicted to be on the slope when the Quinlan court allowed removal of respirators, when the Cruzan court allowed removal of feeding tubes, and when in vitro fertilization helped infertile couples conceive. Predictions of slippery slopes over the past three decades of medical ethics (including some of those made by my fellow medical ethicists) have not been impressive.

If we switch from ethics to the law, the pressing question concerns what kind of evidence we want to allow before we let an incompetent patient die. The *Cruzan* decision said no evidential barrier could be erected by a state for competent patients, that is, whatever a competent patient decides must be accepted, even if it means death. The *Cruzan* decision also said that a state may (not must) adopt a high standard of evidence about allowing incompetent patients to die.

Three general standards of evidence used by courts are (1) preponderance of evidence, (2) clear and convincing, and (3) beyond a reasonable doubt. The weak, first standard had to be met to indict O. J. Simpson, but the most rigorous, last standard must be met to convict him. The clear-and-convincing standard is in between.

Judge Baschab said clear and convincing evidence would be needed to disconnect Correan's feeding tube. That is not what Alabama law currently says. But there is a vacuum in our law.

Alabama's Natural Death Act does not now specifically address feeding tubes, nor does it allow a proxy to be appointed in the event of incompetency, so it does not say what happens when there is no living will to be found. Similarly, although the Alabama statute about durable power of attorney has been interpreted to allow making medical decisions for an incompetent, state law doesn't explicitly specify this the way other states' statutes do. Without such matters being spelled out specifically in state laws, the door is open for third parties and unhappy relatives to challenge a family's decision.

Only two states (New York and Missouri) adopted the standard of clear and convincing evidence for incompetent patients. Generally, this standard means that a formerly competent but now incompetent patient must have signed and had witnessed a living will and kept it on file in a hospital or a safe deposit box.

This standard makes it very tough for families of incompetent patients to cease treatment. It expresses a judgment that allowing for the possibility that a now incompetent patient might have wanted everything done, or that for her to maximize all chances of recovery, it is more important than allowing (and trusting) families to make such decisions for their incompetent relatives. And in my opinion, it wrongly puts the onus of proof on the family to show that the patient would have wanted treatment withdrawn.

Although senior citizens often have written out their wishes for such situations, few younger ones do. Last week, I asked a class of seventy-five students how many had living wills. Three students raised their hands. So if any of the other seventy-two are in a car accident this weekend and end up in a coma like Karen Quinlan and Nancy Cruzan (both young women in their twenties at the time of their accidents), their parents will have no document to meet Judge Baschab's clear-and-convincing standard. They will have to stay on their respirators and feeding tubes indefinitely.

What is bad about this situation is that if physicians or family know they cannot remove a respirator or feeding tube from a patient, it makes them reluctant to use these treatments. But it's good to give patients the chance to see if they will recover. Erecting a high standard about evidence tends to cut off treatment options.

A model bill to reform Alabama's laws about death and dying has languished in Montgomery for years now, awaiting some champion to shepherd it through the channels of our legislative process. It is time for this bill to pass. Otherwise, any of us could end up like Correan Salter.*

---

*About five years after this piece appeared, the Alabama legislature passed a much-improved Natural Death Act, although before that happened, Salter's body collapsed from additional strokes.

21

In Case of Terminal Illness, Call Your Lawyer, Not Your Physician

It has been thirty years since reporter Shana Alexander helped birth bioethics by writing a *Life* magazine article about how Seattle's "God committee" chose which few patients would live by receiving then-scarce dialysis machines. Most of the thousands of then-eligible patients were not selected and soon died.

At a Seattle conference this October [1992] marking the "Birthday of Bioethics," Alexander, who is now approaching retirement age, confessed that she has given a lot of thought lately to issues of dying and hospitals.

"When it comes to following my wishes about how I want to die," she told the large group, "I trust my lawyer more than my physician."

For physicians who see themselves as morally superior to lawyers, her remark hurt. But most people in the lay audience nodded in agreement.

The traditional ethos of medicine encourages physicians to make moral judgments about what is best for patients and to act

---

Originally published as "In Case of Terminal Illness, Call Your Lawyer," *The Tuscaloosa News* (AL), 15 November 1991, 9A.

on those judgments. Unfortunately, sometimes tragic conflict occurs when a strong-willed patient resists that judgment or when there is simply a conflict of moral judgments about how and when that patient should die.

Even if a patient provides an advanced directive about how she wants to die and designates a person to carry out her wishes, another relative may challenge a decision by health care providers to let the patient die. Hospital administrators, as well as physicians, may take steps to avoid suit by that relative.

In many cases, the easiest thing to do is let the dying drag out, even though that may be precisely what the patient wanted to avoid.

In such cases, the bioethicist inside the hospital or the ethics committee who urges following the patient's wishes may encounter the momentum of the institution for self-protection. But tension is inevitable because the physician will be mindful that the physician's practice is on the line for possible suit, not the bioethicist's.

I wish physicians would worry less about lawsuits and reassert their commitment to patient decision-making. I wish the medical profession would publicly assure would-be patients such as Shana Alexander that physicians will do what patients want, no matter what.

Americans want a physician who will say to them, "After we've discussed these matters, I'll follow your wishes exactly, even if I personally disagree with your decision. My job is to do what you want and to respect your trust in me and in the medical profession. I'll keep you free of pain, in control as much as possible, and if it comes to it, I'll help you die peacefully, even if it means taking somewhat active means. Even if I must go to court, even if I must fight my own hospital, even if I must fight your family's lawyers, I will still be on your side."

Of course, this is naive. Of course, it is traumatic to be sued. Of course, each case is different and complicated in its own way.

But what is needed about dying in America today is a passionate commitment by physicians, not to the patient's welfare as a particular physician philosophically sees it, but to reassuring patients that their wishes will be followed.

Too often, physicians are on the wrong side of cases involving patient rights. With a few recent exceptions, most physicians seem unwilling to risk anything to publicly defend patient rights. It has fallen to the much-maligned legal profession to fight for patients' interests.

The Quinlans needed a lawyer to fight the physicians and the hospital who opposed their wishes for Karen.

Elizabeth Bouvia needed a lawyer to resist the head physician who said, "Because she's occupying our space, she must do as we say."

The Tuskegee victims in Alabama needed a lawyer after they had mistakenly trusted physicians of the federal government who deliberately omitted giving them penicillin for thirty years for their syphilis.

The homeless New York City woman Joyce Brown needed lawyers to prevent her from being involuntarily committed by Bellevue Hospital physicians.

In Alabama, the Medical Association rammed a law through the legislature allowing any physician or dentist to test anyone for HIV infection, without the patient's consent, largely in order to protect surgeons and other physicians against perceived risk.

Whose interests were the physicians defending?

Why are physicians so often on the wrong side?

I believe that Americans want physicians to be on their side.

I wish physicians would respond.

Recently, I addressed a suburban book club, all of whose members were affluent females aged forty-five to eighty-eight. About half were married to physicians in Birmingham.

Even among this elite group, all the women feared that, once inside a hospital, they would lose control of their lives. Many had already experienced terrible problems in dealing with physicians and hospitals with their own elderly parents. All of these women agreed with Shana Alexander's remark that they trusted lawyers more than physicians to do what they wanted.

I hope American physicians are listening to these women.

22

Everyone Creates Soaring Medical Costs

I was once a liberal about financing medical care, then I became a conservative, now I am a nothing. I once thought physicians' fees caused increases in medical costs, then I blamed large institutions; now I blame everybody and nobody. I once thought escalating medical costs were a problem, later I thought they were a crisis; now I think our system may collapse before the second millennium.

I came to medicine as a new philosophy Ph.D. who had written a thesis about Rawls's liberal theory of justice. Starting in 1977, I began making rounds with physicians, seeing patients in oncology services, dialysis units, and public clinics. Since then, I've talked with many physicians, learned much about what medicine can and cannot do, and come to believe that ominous patterns are emerging in American life and medical care.

Some people may doubt my belief that medical costs are uncontrollable. After all, Social Security was saved recently and surely we can do the same for soaring medical costs. What the critics

---

Originally published as "Everyone's Entitled to Blame for Soaring Health Costs," *Wall Street Journal*, 22 December 1983, A25.

pointed out and what the Social Security Commission purposely overlooked was that three-fourths of Social Security's problems are in rising medical costs. The commission did nothing about this problem, and Social Security, as well as medical finance, is still in trouble.

Consider some alarming trends. Between 1968 and 1980, medical payments for veterans jumped 410 percent to $6.6 billion from $1.6 billion (after which two presidents vastly increased military personnel and their benefits). Furthermore, and for various reasons, between 1960 and 1980, recipients of Aid to Families With Dependent Children (AFDC) jumped from 3 million to 11 million with these youngsters entering the world on Medicaid.

Physicians, too, must bear some blame. Very soon there will be too many physicians, and a glut of physicians creates not competition but overuse of medical services. Moreover, just as veterans, AFDC recipients, and middle-class patients feel entitled to their benefits, so physicians feel, after a decade of sacrifice and an average debt at graduation in 1982 of $28,000 [in 2002, the average debt at graduation is $68,000], entitled to a salary often exceeding $100,000 a year.

Many other factors contribute to runaway medical costs, each bearing its own entitlement. Private companies, managing hospitals and nursing homes or making pacemakers and dialysis equipment, feel entitled to a good profit (like the oldest for-profit medical business—pharmaceuticals). For every physician, there are seven allied health professionals, each wanting more pay and status, together bloating the national health-care bill. (There are even philosophers on the payroll at medical schools now!)

Hospitals finance expansions through interest-bearing bonds, building inflation into future medical costs. More ominously, we may have a permanent class of unemployed and underemployed workers who do not qualify for Medicaid but who lack private medical insurance. It is unlikely that we are just going to let these people die when they present themselves to emergency rooms late at night, and since they have the worst health, the cost of their care will be high.

Lest we blame everyone but ourselves, notice, too, how we've become accustomed to seeing medical insurance as paying for most of our medical care instead of insuring us against catastrophic illness and accident. If our policies cover dentistry, eyeglasses, or even psychotherapy, we now feel entitled to these services.

The most ominous factor of all behind the coming crisis in medical finance concerns something of which I am a part: the baby-boom generation. Sometime in the next few decades, my friends and I will awake to find ourselves suffering from gallbladder problems, kidney stones, and heart murmurs, and we will begin to knock hard at medicine's door. We will say, "We've paid for all these years; now we deserve something back!" At the same time, the marvelous advances in medicine (as well as its retreats, in maintaining indefinitely bodies whose lives are over) will bequeath to us a large, medically expensive population made up of our parents. I feel frightened in realizing that the workers who will be obligated to pay for all the medical care for the by-then majority of senior citizens will be our children. The burden may be intolerable.

I've stressed the word "entitled" so far to buttress my conclusion: Medical costs are uncontrollable because we lack moral agreement about how to deny medical services. In fact, we've inherited incompatible, contradictory systems: in justice, a raw property-rights conservatism, for which high taxes are a kind of slavery, running up against Judeo-Christian egalitarianism, in which the effects of Fate's unequal health lottery must be minimized by a just society. Here claims for more personal income and lower taxes fight those of institutionalized charity. In ethics, we've inherited a sanctity-of-life principle from Hippocrates (a pagan who followed the mystic Pythagoras), as well as the dominant quality-of-life principle of Socrates and most ancient Greek physicians.

Combinations and permutations of these four beliefs can be interesting: low-taxation champions who favor government bans on abortion and forced, expensive treatment of defective babies; angry physicians complaining about government interference who murmur no protest when unions force businesses to pay for exorbitant medical policies or when governments "interfere" with workers to pay for medical care by increasing taxes.

Alas! Few of us are totally consistent in such emotionally charged areas. We want lower medical costs, but we want everyone to be treated.

Technology and medical advances have intensified, not solved, such moral dilemmas: A hemophiliac lacking a clotting factor does not die today, but his Autoplex units for three months cost $100,000; the usual method for feeding children who lack a digestive tract costs even more for a year of an infant's life but dooms the child to live in a hospital; the fantastically expensive neonatal intensive care units have saved thousands of preemies and other compromised babies who otherwise would have died and didn't exist thirty years ago. Even inside medicine there are dark rumblings about the new technologies, as seen in a family practice physician's remark to me that "transplant surgeons will bankrupt the system" (followed by a diatribe on surgeons' fees).

Liberals want to subsidize medical care for the medically unfortunate and my compassion agrees with them. The healthy and wealthy should subsidize the sick and poor. Conservatives counter that this cannot go on forever; there are too many people, too many possible medical services, too many entitled, and not enough "obligated." Again, I agree. Moreover, if the system does collapse, I can see that the poor and sick will be hurt the most.

But knowing the desperate need to cut costs, I also understand the physician's dilemma. I have been with him as he tries to tell a teenage girl, in effect, that she's going to die because there's no money. Deciding how to say "no" to such people, and to say it with honesty and integrity, is perhaps the most profound, most difficult moral question our society will face in coming years. But face it we must, for the alternative is disastrous.

23

How to Say "No More" to Patients

The Clinton administration holds out the promise of lower medical costs, universal access, in-home care, nursing home coverage, and a continued private insurance industry. But it has played down the obvious need for a system of rationing expensive treatment. In doing so, it has left a painful truth unsaid: Rationing is not just an economic issue. Rationing is also an ethics issue. In reforming medicine, we must address how physicians will ethically deny the best medical resources to some patients.

Should unrepentant alcoholics get quarter-million-dollar liver transplants? Or, in a society that considers alcoholism a disease, would it be discrimination to deny them a transplant?

Consider patients who get two heart transplants. On average, they live for a shorter time after surgery than those receiving only one transplant. Is it patient abandonment for surgeons not to give patients a second, or even a third, heart?

What about hemophiliacs who prefer a commercial clotting factor that is perfectly free of the HIV virus but costs about

---

Originally published as "Who Gets Health Care? Make a List," *New York Times,* 14 June 1993, A14.

$46,000 per month, versus the 99.9 percent HIV-free clotting factor that costs only $400 a month? Or heart attack victims who prefer tissue plasminogen activator at $2,400 a dose because of its small medical advantage over streptokinase, which costs $200 a dose?

Any health care reforms will be incomplete unless these questions are dealt with and unless we have an answer for physicians when they inevitably ask, "How can I ethically deny treatment to my patients?"

Setting limits will take courage, like that shown by Richard Lamm, the former governor of Colorado, in 1987 when he criticized surgeon Thomas Starzl for performing a $240,000 liver transplant on a seventy-six-year-old woman. Governor Lamm wrote, "It is outrageous that this country spends five to eight times what other countries spend and yet has no better health outcome." He proposed that medical costs be a fixed pie, where a bigger slice for transplants means a smaller slice for something else.

According to April's *New England Journal of Medicine,* only 25 to 30 percent of medical costs for Medicare patients stem from care during the last year of life—a percentage that has not changed over the last decade. But Daniel Callahan of the Hastings Center, a medical ethics think-tank in Briarcliff Manor, New York, says 10 percent of patients account for 75 percent of all medical costs. So big savings won't come from foregoing expensive technology in the last year of life but from denying treatment to a few people with extremely expensive problems.

What do we tell the seventy-six-year-old woman with liver failure? The easy way out is to never discuss a transplant with her, as has been suggested by some health care planners.

But nondisclosure is cowardly. When you fire someone, you do it yourself and you explain why. It's not easy, but it's the right thing to do. Obviously it is much harder to give bad news face-to-face when life itself is at stake, but it's the right thing to do. Can Americans meet the ethics of rationing face-to-face? The behind-closed-doors approach of Hillary Clinton's health care task force has hindered reaching a national, moral consensus. But it is possible to follow the example of Oregon, which had town meetings and created a reasonable rationing plan that was approved in 1993. The Oregon plan was almost derailed by charges that it discriminated against the disabled. We must listen to all voices in the search for a rationing plan, but we cannot let the first cry of discrimination put the effort in jeopardy.

To the seventy-six-year-old woman with liver failure, we must say: "For all your children and grandchildren, we can't spend this much on you." To the patient with one heart transplant: "I'm sorry, but we can't afford to give you more than one heart because it costs too much and because another person awaits the next heart." To lifelong smokers: "Sorry, no lung transplants. You could have stopped smoking."

If we don't start publicly talking about rationing, we'll never get a consensus and medical reform will fail.

24

What the Clinton Medical Plan Should Have Emphasized

Dear Mr. President and First Lady,

The philosopher Wittgenstein once wrote that in any real philosophical debate, the first step of how the debate is set up often goes unnoticed. How the parameters are defined, how the criteria are determined, how the media frames an issue—these steps are often not discussed.

In our current debate in the national media about medical care, the issue has been defined almost exclusively in terms of finance. I think that is a mistake. If we only look at reforming medical care this way, we lose the big picture.

Now that you've outlined your plan, you must wrestle with how this debate should continue.

Because we Americans often set our opinions in stone, Mr. President, you should consider what you want the country to focus on: financing access to medical care or whether it's right to include everyone in our medical system.

We have been asked many times to compare costs of medical care in Canada with those in America; to ponder the dilemma of

---

Originally published as "An Rx for Health Care?" *Birmingham News,* 26 September 1993.

whether to pay for greater care by taxing business, all Americans, or all patients; and to consider whether savings from various reforms can pay for the costs of covering uninsured Americans.

I applaud you and Mrs. Clinton for your desire to obtain basic medical care for all Americans. Your promise to do so was the reason that tens of millions of Americans voted for you. But, if it's not too presumptuous of me, I want to offer you some advice.

Obsession about the financing of care misses the major impetus behind the movement to change American medicine: the issue of justice.

I think, Mr. President, that you, Mrs. Clinton, and your staff would be much more successful in rallying Americans behind you if you kept the focus on the morality of a country as wealthy as ours not having basic medical care for 32 million of our brethren.

We Americans will respond to a clear appeal to our hearts. We get bored with details of financial plans. Put the question to us as one of compassion and justice, not of HMOs and managed care plans.

I am well aware that people define morality and justice in different ways, but one basic, central notion of justice underlies all of our differences. Perhaps the most influential concept of justice of this century stems from the justice-as-fairness conception of Harvard philosopher John Rawls.

To determine the most fundamental principles of justice, Rawls asks us to imagine ourselves in a social contract where a "veil of ignorance" falls over our personal characteristics such as race, gender, age, class, income, family size, or religion. Under such conditions of ignorance, Rawls believes that we would choose principles of equal liberty for all and fair equality of opportunity for everyone.

In essence Rawls's conception forces people to choose the basic structure of society under the Golden Rule.

The principles chosen apply to the basic structure of society and will be much broader than U.S. Supreme Court decisions and far broader than specific legal decisions. Very close to this deepest level (perhaps one level above it) is the choice of the basic structure of the medical system.

Critics will immediately object from both sides: reformers that we have no system; conservatives that no one has ever really chosen the present status quo in medicine.

That is precisely the point. We need an ideal of justice to guide us, so we can see whether we are going in the right direction. Without such an ideal, we will be just as aimless as in the past and arrive at a similar or worse mess.

What would Americans choose under Rawls's veil of ignorance? I believe we would choose to give everyone access to basic medical care.

Why? For one thing, it would be rational to do so. When the veil rises we do not know whether we will be a teenage mother with three children on public assistance or Donald Trump. Most of us would not bet on being Donald Trump.

Moreover, Americans would choose universal access as in Canada because lack of access destroys lives. Any conception of the good life requires access to care for illness and injury. When we get sick or injured, we become scared and panicked. It is almost unimaginable to me how scared and panicked people who have no insurance at all must become when they are ill or injured. And imagine being one of the 17 percent of American women who give birth, a scary enough experience in and of itself, with no medical insurance.

You, Hillary, were correct when you stressed that the real issue was security about medical care and "peace of mind." Despite its critics, Medicare has been a great success in providing reliable, good medical care for those over sixty-five. Surely we owe the same to the sick and infirm of all ages.

More and more, scientific research is revealing that whether we get breast cancer, heart disease, or mental illness depends on the random spin of the genetic roulette wheel when the chromosomes of our parents mix during conception. More and more, science is falsifying the current great ideology of our times: that healthy living can prevent all disease and, therefore, that those who get sick did not live correctly. The sad truth more and more is that it's often just bad luck when you get brain cancer at thirty-six.

Consider all the children. Many of the uninsured today in America are children. To not vaccinate children against childhood diseases, to not pay to have broken bones set correctly, to allow children to die from untreated meningitis is unpardonable. Who would vote in the social contract for a medical system where his or her children could be stricken with cancer and there would be no money for treatment?

The great problem of focusing of finance is that it puts us on the wrong path at a crucial fork in the road. Is our problem mainly one of justice and then of how to pay for creating a just medical system? Or is it chiefly a financial one with some moral implications?

I think reforming the medical system must be addressed as an issue of justice.

Consider how biased the discussion of medical insurance has become by focusing on finance and on the survival of the estimated 1,500 private companies that sell medical insurance. Medical insurance can be conceptualized in two broad ways: first, as a moral endeavor in which we all pay small premiums to insure ourselves against an unlikely accident or sudden illness, and second as just another way to make profits, no different than selling cars or real estate.

The moral conception of insurance roughly corresponds to community-based policies where companies were required to insure everyone in a "community," such as the state of Tennessee, at the same rate. In the past three decades, changes in laws allowed private companies to sabotage this moral conception and to cherry-pick the most profitable clients by using experience-based policies, selling policies to the young and healthy and denying them to the old and sick.

If we don't see the history of our current problem, and see it only as one of whom to tax, we miss a lot. Consider also that the parents and grandparents of those without insurance today paid taxes for the public hospitals, medical schools, and U.S. Public Health Service that benefit all Americans. Those working without insurance today also pay the FICA taxes that finance the revolving-door funding of Medicare, yet they receive no similar security.

Mr. and Mrs. Clinton, I know we can finance basic medical care for all Americans. To me it isn't important whether we pay for it by doing fewer transplants, eliminating private insurance companies for a single-payer Medicare-type system, making people pay higher copayments, or quadrupling taxes on cigarettes and alcohol. That's the fine print.

What you need to hold up before the American people, like President Reagan's "City on a Hill," is the image of a nation where a poor American child's life will not be destroyed because of lack of good medical care or because we Americans are too cheap, too selfish, too misinformed, or too manipulated by slick profiteers within the medical-industrial complex to create such a system.

Don't let us focus too heavily on whether we pay a few more dollars for a prescription (that we may not have gotten if we had to pay for it entirely ourselves). Keep us thinking about what it would feel like, once the veil rises, to be the kid with cancer without medical insurance. Do that and lead us toward really being that shining city on the hill.

25

Should Doctors Treat People with AIDS?

No other issue in modern medical ethics has the clout of AIDS. It dominates ethics classes from medical schools to high schools, and issues concerning the ethics of how to treat AIDS patients haunt health professionals today. Most of these health care professionals accidentally get stuck by needles carrying patients' blood at some point, even though they work carefully, and most have to care for an HIV-infected person at one time or another.

In the November 27 [1992] issue of the *Journal of the American Medical Association* (*JAMA*), results of a survey were published that surprised some people. A survey of two thousand physicians on the front lines of health care revealed that, if they had a choice, half would not treat people with the HIV virus, and nearly half preferred to refer such patients to other, allegedly better qualified, physicians.

Doctors were soundly trashed for these results, especially by two physicians who wrote an accompanying editorial. However,

---

Originally published as "Ethics and AIDS," *Birmingham News*, 2 February 1992, C1.

people may have misrepresented these data: more than 80 percent of the surveyed physicians said that they believed they did, in fact, have a moral responsibility to treat HIV-infected patients.

That is good news. But these physicians also said, quite honestly, that if they had their choice, they would rather not treat HIV-infected patients. Sure, they said, we would all prefer to let someone else handle AIDS.

But what is missing is the fact that, even when they don't want to, most physicians fulfill their moral responsibilities. Many a young physician has wanted to go home and sleep rather than check on a patient, but medicine is a career of increasing commitments to doing what is good for patients, not what is good for oneself. That fact has wreaked havoc on the personal and family lives of many doctors.

Medicine, moreover, has already established that the profession is going to care for people with AIDS. Surgeon General C. Everett Koop lashed out in 1987 at physicians who declined to treat HIV-infected patients.

When John Ring, the president of the American Medical Association, was in Birmingham two weeks ago for a conference about ethics, he said that the basis of professionalism in medicine is "competence, ethics, and compassion." On this basis alone, all competent physicians have an obligation to treat HIV-infected patients.

Some physicians prefer to see medical degrees and licenses as proprietary, not something bestowed by the public to ensure the best care for the sick. For them, their personal liberty rights to treat whom they choose, bill as they choose, and work where they choose are no different than those of any other businessperson.

George Lundberg, editor of *JAMA* and a University of Alabama at Birmingham (UAB) Medical School graduate, was also in Birmingham for the same conference, and he asked, "Are we in a profession to which business is incidental or in a business where professionalism is incidental?" So professional physicians are going to see to it that HIV-infected patients get treated.

At UAB, which trains most of Alabama's physicians and which is the state's largest employer, it is an official policy that a student, nurse, or physician must take care of HIV-infected patients who come under his or her normal duties. There is no right to not treat such patients.

Imagine if there were. It would mean that a few heroic physicians would treat all the HIV-infected patients. Indeed, the problem of which physicians treat HIV-infected patients is largely a problem among doctors, internal to the profession. Ditto in dentistry or any subspecialty. When a local physician tells an HIV-infected patient that he is "not competent" to treat AIDS and refers the patient to UAB, word gets around quickly at UAB and in the county medical association about the doctor's attitude. (If the doctor is really not competent, why not? Physicians have an obligation to keep up with their fields or retire.)

It is perfectly obvious to everyone that if we cut away the many add-on ideas about what physicians should do—whether they should be administrators, teachers, activists, researchers, grant writers, scientists, novelists, Renaissance men, or commentators on television—we finally hit one rock-bottom activity that defines a doctor: one human being helping another human being cope with illness. The integrity of that physician-patient relationship carries the weight of everything else. If there is no integrity in that relationship, the edifice above crumbles.

On the other hand, the public has not given the health professions much credit for the fact that most physicians, nurses, and dentists struggle daily with problems concerning AIDS that would terrify ordinary people.

There has been so much irrational publicity about the remote chance of accidentally becoming infected with HIV from an HIV-infected surgeon or dentist that the real heroism of those caring for people with AIDS has been consistently overlooked. Most of us underestimate how, if we were the one responsible for the care, we ourselves might fear getting the HIV virus from a patient. Consider Abagail Zuger, once a young, highly ethical resident in New York's Bellevue Hospital, writing in the *Hastings Center Report* about her residency back in 1988 when there were many fewer cases of AIDS:

> In my three years at Bellevue, AIDS came to engulf the hospital; it dominated my medical training and now it dominates my memories. I remember the weekend when no patient in the intensive care unit was over 40. I remember the intern who tearfully refused to come to the emergency room to see the fourth AIDS patient I had admitted to her in as many hours. She never did meet him; he died before she calmed down.
>
> I remember the meal trays, stacked precisely three and four deep, undisturbed, outside the closed doors of the private rooms of one of the medical wards. And I remember Nilda, a drug addict who, like many others at the time, had a fever we couldn't explain; one morning, in my usual state of exhaustion, I jammed the needle I had just used to draw her blood deep into my thumb.

A typical resident in internal medicine in New York City cared for thirty HIV-infected patients during 1988 and stuck himself or herself 1.3 times with known HIV-positive blood. Today, the AIDS caseload and number of needle-sticks there have probably doubled. (The same issue of *JAMA* reported a nearly 30 percent drop between 1983 and 1990 in medical school graduates who wanted to do their residency in New York City.)

Of course, being stuck with HIV-positive blood doesn't mean you're going to get infected yourself. The chance is very, very small. But that doesn't make it any easier to take the test every few months. And should the resident tell his wife about the needle-stick? What if she is AIDS-phobic and refuses to have sex until he tests negative for six months? It has happened. But if he doesn't tell her and has the terrible luck to become infected, he runs the risk of giving her the lethal virus. And what if she is pregnant? Does he want it on his conscience that he might have helped kill his wife and child by not refraining from sex after his needle-stick?

If this sounds crazy, talk to people who have actually been stuck with HIV-infected blood and who nervously await their results.

Finally, any honest discussion about ethics and AIDS has to deal with people's attitudes toward homosexuality. In the above *JAMA* survey, an astonishing 35 percent of physicians said they "would feel nervous" simply being among a group of gay people and the same 35 percent agreed with the statement that "homosexuality is a threat to many of our basic social institutions."

Molly Cooke, a medical ethicist at San Francisco General Hospital, found that this attitude is found more in male physicians toward gay men than toward lesbians or among female physicians.

In the medical school ethics course I teach, I always have one class meeting at which gay people with HIV infections participate in a panel and answer questions from medical students. Some of the questions are hateful and ignorant. I can't help wondering how these underlying attitudes carry over into the students' work in the hospital, especially their work with HIV-infected patients.

Of course, one small course in ethics does little to change attitudes on such an intense subject. Still, it's helpful to hear both sides argued, especially in discussions with actual gay people living with HIV infections. For this reason, it is shocking that many medical (and almost all dental) schools still lack an ethics course where such issues are discussed. That is a kind of educational malpractice.

26

How Politicization of Facts about AIDS Helped Kill People with AIDS

Since gay men began coming down with a mysterious disease in the summer of 1981, facts about AIDS have been controversial. The disease over the years has been associated with homosexuality, death, sex, intravenous drug usage, and disfiguration. From the start, the disease raised humanity's worse fears. Each person has seen facts about AIDS from his or her own perspective. Humans often allow their prejudices, ambitions, and fears to interfere with their judgment of facts on controversial issues. Despite their training, scientists and physicians are also prone to this vice.

Many past assumptions in medicine distorted factual judgments. In the antebellum South, slavery was partially rationalized by the belief that blacks were biologically different and racially inferior and, hence, medicine held that pathology in blacks had to differ from in whites. In nineteenth-century America, sexual desire in females was equated with pathological promiscuity and was sometimes cured by ovariotomies and clitorectomies. Cholera was

---

Originally published in a longer form as "Evaluative Assumptions and Facts about AIDS," *Ethics and AIDS: Biomedical Reviews*, 1988, ed. R. Almeder and J. Humber (Clifton, NJ: Humanities Press, 1989), 5–19.

blamed on the vices of drunk Irish, lazy blacks, wanton prostitutes, and the dirty poor, but not on infected water. One nineteenth-century minister claimed that God sent it "to rid the filth and scum from the earth." Similarly, so tight was the conceptual bond between sin and syphilis that early-twentieth-century physicians resisted evidence that spirochetes caused syphilis.

Transmission of such diseases by germs did not fit the evaluative paradigms of the times. In all these cases, what counted as medical fact was what blamed the victim for the condition.

History of Controversial Facts about AIDS

Many facts about AIDS are more in dispute than facts about other diseases because of the nature of AIDS. If an evil demon had tried to give humans a disease that would bring out their greatest irrationalities, he could have done no better than to create one that: (1) is lethal; (2) makes its young victims die in horrible ways; (3) is transmitted by sex; (4) is transmitted by unapproved forms of sex; (5) afflicts victims who are already stigmatized and, hence, gives any new victims its stigma; and (6) has mysterious origins, transmission, incubation, incidence, and development.

All facts about AIDS exist in a social human context. Most factual claims to date about AIDS have been fraught with distorting biases. Caution is necessary to prevent humans from taking their own expectations, hopes, and fears as facts. Because many different kinds of people care how things turn out, establishing even the most elementary fact has been difficult. Every time one scientist makes a claim, another scientist seems to dispute it.

Speaking about AIDS

A complicating factor in determining the facts about AIDS is the different ways in which people speak in public. Scientists tend to report the evidence and let their peers draw their own conclusions. Some scientists encourage relaying such reports to the general public, arguing democratically that most people can evaluate evidence and statistics. Others argue that public spokespersons must arrange and filter scientific evidence, pronouncing simple guidelines that even illiterate teenagers can understand. Some anti-AIDS crusaders criticize scientists for not making absolutist prohibitions in the graphic language of teenagers. Scientists reply that the present, probabilistic knowledge about AIDS, as well as precise scientific language, is not compatible with simplistic generalizations.

Because such low-income, often illiterate teenagers are perceived to be most at risk for AIDS, a paradox exists in AIDS talk. Communications with such teenagers as targets often appear in newspapers and magazines, yet those teenagers most at risk seldom read. Those who read the most, on the other hand, are likely to overreact because of inundation.

The First Victims of AIDS

The identities of the first victims of AIDS were crucial in creating later controversies. When AIDS first appeared in the early 1980s, fundamentalists like Jerry Falwell, as well as representatives of the Catholic, Baptist, and Mormon churches, denounced homosexuality and said that God was punishing sin with AIDS. Condemnation increased when IV-drug users were added to the list of victims of AIDS. Social conservatives jumped on board, and Patrick Buchanan said, "The poor homosexuals—they have declared war on nature, and now nature is exacting an awful retribution."

Certain assumptions were initially made in medicine about AIDS victims, like similar assumptions about past victims of disease, and these initially determined what we believed to be the facts about AIDS. Thus the early medical assumption was made that only gays got AIDS, and this implied that something about gay sex transmitted HIV. This assumption blinded the American Red

Cross and its medical advisors to the possibility that HIV was transmitted in the blood supply. Intravenous drug users and their sexual partners similarly perceived little risk.

How AIDS Facts Were Reported in the Media

Although most Americans were prejudiced against gay men, they did not hate gay men. They laughed at AIDS jokes, but believed that gay sex should not be illegal. They were emotionally negative toward homosexuality, but intellectually tolerant. In effect, they were rationalists in the streets and moralists in the sheets. So rational Americans did not condemn gay AIDS patients and did not believe they deserved to die.

But then in the early 1980s, liberal journalists counterattacked prejudicial attitudes. Liberal academics struck a tone alternating between cool rationality in chiding gay-haters and lofty indignation at the media's sensationalization of the topic, implying that AIDS was neither fatal nor a disease of gay males. Liberal clergy urged compassion, and the American Civil Liberties Union opposed restricting liberties of AIDS patients for public health.

Newspapers, television networks, and magazines struggled over how to report AIDS. Unscrupulous magazines and tabloids implied that AIDS was contagious, that anyone could get AIDS, and that angry gay AIDS patients were purposefully infecting the blood supply. As a result, most journalists were wary of being used by bigots. Moreover, most of these reporters and journalists had little training in science and were reluctant to report bad news before scientists achieved consensus.

These attacks and counterattacks paralyzed Americans. Reporters and physicians alike wanted to be seen as neither bigots nor Pollyannas. Every fact had such emotionally charged implications for sexual freedom, for reinforcing prejudice, and for public health that no one wanted to pass on new claims unless they were certain facts. And there were few certain facts, even though there were several experts talking as if there were.

Controversy also developed within medicine as to whether Robert Gallo of the National Institutes of Health had codiscovered HIV or had stolen credit for the discovery from the Frenchman Luc Montagnier. Accounts of rivalries between the two men reveal that facts about AIDS were distorted to soothe Gallo's ego, bolster chances of his winning a Nobel Prize, and protect American medical chauvinism. Even the name of HIV, "Human Immunodeficiency Virus," was for years controversial because Gallo wanted a different virus that he alone had discovered to be the real cause of AIDS.

Randy Shilts, author of the great book about AIDS *And the Band Played On* (1987), as well as leading scientists, publicly questioned Gallo's integrity. On the other hand, *American Medical News* referred to Gallo (as others did) as "America's leading AIDS researcher." Who could Americans believe? Who had the facts?

Personal biases also distorted facts about the transmission of AIDS. In 1982, it was hypothesized that the AIDS microbe was only transmitted by receptive anal sex with multiple partners. Heterosexuals wanted to believe that heterosexuals were safe, especially if they did not have receptive anal sex with bisexual men, and they wanted to believe that AIDS was a gay disease. Likewise, gays wanted AIDS to be an isolated, rare disease, not easily transmissible via sex.

An unsettling example was how gay men themselves adamantly resisted evidence about AIDS, even though such resistance threatened and shortened their lives. By the end of 1985, more than 19,000 cases of AIDS had been reported to the Centers for Disease Control (CDC), and more than 8,000 had died, yet gay periodicals underreported the problem or ignored it.

Academics were little better. In 1986, Jonathan Lieberson still stridently insisted in a lengthy, passionate article in the *New York Review of Books* (January 26, 1986) that having antibodies to HIV didn't necessarily mean infection by HIV, and that infection by HIV didn't necessarily mean getting ARC (AIDS-Related Complex) or AIDS. Lieberson also happily quoted a French physician named Leibovitch who predicted that less than 10 percent of HIV-infected people would experience "any symptoms at all," and that most people with ARC would not get AIDS.

During 1984 and 1985, two conservative views clashed: (1) AIDS was a disease of a deviant minority, and (2) AIDS was a disease anyone could get (as a headline blared in a famous *LIFE* cover). These two views continued to clash for the next four years. The first implied it was not John Doe's worry, and the second that it was. If AIDS was only a gay disease, callous heterosexuals could ignore the problem, but if universal infection was possible, callous heterosexuals should tattoo or quarantine the infected. Both views were countered by scientists who said that no certain evidence existed about anything about AIDS.

As recounted by Shilts, evidence in 1984 was mounting that HIV could be transmitted by blood transfusions. Evidence for this was seen as hemophiliacs and babies of female drug addicts contracted AIDS. Yet so strong was the counterforce of liberal tolerance against the perceived bigoted hysteria of homophobes that obvious evidence was ignored. The Reagan administration, intent on cutting costs, sided with blood bankers. Gays themselves strongly resisted mounting evidence of blood transmission.

In 1983, public health authorities such as Health and Human Services (HHS) secretary Heckler and physician Joseph Bove of the blood-safety committee of the Food and Drug Administration (FDA) erroneously assured Americans that they would not get AIDS from blood transfusions. In 1984 and before the ELISA test began to screen blood in February 1985, donated blood could have been screened for hepatitis B and doing so might have removed perhaps 80 percent of HIV-positive blood. Because the blood was not screened, perhaps thirty thousand Americans got infected. This was a steep price for Americans to pay for not being able to agree on the facts.

Personal desires also entered the controversy over whether HIV was casually transmissible. Fears about casual transmission of AIDS, that is, transmission in ways other than through bodily fluids, grew in 1983 because the facts were unclear. Phrases such as "sexually active," "casual transmission," and "bodily fluids" were vague.

Again, vested interests hoped medicine would find certain facts and not others. The American Medical Association (AMA) released an erroneous press release from an erroneous editorial reporting eight "unexplained" cases of AIDS in children, presumably by household contact. Shortly thereafter, fears began about HIV-infected children in schools. Parents worried about an HIV-positive child biting another and were told it couldn't happen. Not too long thereafter, HIV was cultured from the saliva of an HIV-infected patient. In 1988, this saliva study was retracted.

More Uncertainties

Mistakes about gays and AIDS are easy to see in retrospect. Such mistakes predictably spilled over to factual claims about risk to medical workers, heterosexual transmission, and prevention. Medical workers were told in the early 1980s that HIV could not be transmitted through a needle stick. When some of them became HIV positive, it was implied that they had a secret life. As of the beginning of 1988, at least fourteen medical workers had become HIV positive from occupational accidents. This figure was not well publicized and, some claimed, may have been kept quiet to stem the present flight of nurses and residents from high–AIDS caseload hospitals. By 1988, the claim about secret lives had been largely dropped.

Medical workers were also told in the early 1980s that HIV could not be spread by contact with mucous membranes, and then three cases were reported. Those desiring that such contact not be possible also implied, at first, that the infected workers had a secret life. This claim was soon also dropped.

Medical residents in four cities have cared for 50 percent of AIDS patients and performed perhaps 75 percent of physician's duties toward AIDS patients. Such residents, especially after a thirty-six-hour shift with no sleep, usually stuck themselves with HIV-infected blood at least once a year. Given previous mistakes about facts, they were understandably scared, and some objected that those taking the high moral road on treating AIDS patients had rarely themselves treated AIDS patients, much less thirty patients a year, much less year after year, much less after many needle sticks.

While telling patients and nurses that their chances of contracting AIDS from medical care were small, other physicians' actions were telling people different things. Surgeons demanded that patients be HIV-tested before surgery and claimed they were at risk.

In San Francisco General Hospital, a leading hospital for treatment of AIDS patients, orthopedic surgeons demanded such testing after a worker seroconverted (became HIV positive) after a needle stick. Surgeon Loraine Day knew that only a 1 in 800 theoretical risk existed of contracting the virus from a single needle stick, but she operated on many gay patients and constantly cut herself. "I may get stuck 20 times in the next six months," she said, "which means my risk is now 1/40. Over the next year, my risk is 1/20. This is not a low risk, it's very high risk." Day also accuses physicians of what Randy Shilts calls "AIDS Speak" (*American Medical News*, 4 December 1987). Her superiors talk of the worker who "seroconverted," but Day says, "Let's be honest. Let's say the hospital worker has contracted a terminal illness and will die."

Other hospitals and plastic surgeons across the country quietly instituted patient testing, leaving residents with an uncertain message.

Heterosexuals and AIDS

The same factual uncertainty predictably appeared about heterosexual transmission of HIV. In early 1983, many scientists believed that people could not become infected through heterosexual sex. This belief changed over the years.

Concern over heterosexual transmission grew feverish in early 1988. At the beginning of 1987, HHS secretary Otis Bowen warned of the imminent spread of HIV to the heterosexual population and urged heterosexuals to refrain from unsafe sex. Just a year later, he announced that the much-feared spread to heterosexuals had not occurred and probably never would. America's other top health official, Surgeon General Koop, vehemently disagreed with Bowen. During February 1988, psychiatrist Robert Gould wrote in *Cosmopolitan* magazine that "there is almost no danger of contracting AIDS through ordinary sexual intercourse. . . ."

A month later, noted sex researchers Masters and Johnson threw gasoline on the fire, claiming that "AIDS is now running rampant in the heterosexual population," that HIV could conceivably be transmitted by kissing and on toilet seats, and that 6 percent (twenty-four people) of four hundred sexually active heterosexuals (more than six partners in a year, each year, for five years) were found by them to be HIV positive (*Crisis: Heterosexual Behavior in the Age of AIDS*, Grove Press, 1988). *Newsweek* blared the story on its cover and excerpted the Masters and Johnson book (*Time* refused the opportunity). AIDS researcher Matilde Krim criticized Masters and Johnson for their alarmist calls, for their "science by press conference," and for their eagerness to sell their book.

In the larger public, heterosexuals did not feel at risk. In this feeling of safety, feelings of risk stemmed from perceptions not about the disease but about the nature of the victims: because I am not one of "them," I am not at risk. Because I do not do "that," I am safe.

What counted as a fact about how AIDS was contracted was determined by perceptions about who got AIDS, not about the evidence itself. Masters and Johnson emphasized that in past sexually transmitted diseases, such as syphilis and herpes, 5 percent of the people accounted for 50 percent of new cases. (This was the target group of their nonrandom study.) Given their results, the most dangerous infected heterosexuals probably looked perfectly clean and safe to prospective partners.

Fundamentalists and public health professors argued that HIV was highly contagious in any kind of sex and that it was infecting everyone. "Anyone can get AIDS," they taught high school students. They pointed to Africa, where HIV appeared to spread heterosexually among Africans. Heterosexuals hoping for a reprieve argued that Africans were "other," and that Africans spread HIV through genital ulcers, religious rites, and reusing needles. Heterosexual men argued that evidence was scant that HIV could be transmitted in normal intercourse from women to men. It is obvious what they wanted to believe.

Condoms

The efficacy of condoms in preventing the spread of HIV was also factually disputed. Social conservatives cited FDA tests showing a 20 percent failure rate and argued that condoms often failed to prevent pregnancy (emphasizing that women could only conceive for a few days during a month). They said a virus was much smaller than a bacterium and capable of penetrating latex interstices. Their solution was abstinence or partner fidelity. In this subjectivist paradigm, it was difficult to determine if their solution was an inference from the facts or an assumption determining how the facts were seen.

On the other hand, single heterosexuals pinned great hopes on the safety of condoms. Here evidence was especially murky as to whether condoms reduced or eliminated risk. Getting HIV from infected blood splashed on a hangnail sore was very unlikely, but that it could happen at all scared some medical workers. If a similar chance existed of acquiring HIV while using condoms, that fact would be most unwelcome in certain quarters.

Animal Models

Another hotly disputed fact was whether chimpanzees should be used in AIDS research. Robert Gallo found them indispensable and lamented the strength of the animal-rights lobby. Physicians for Social Responsibility and animal-rights champions said chimpanzees were neither necessary nor good models for testing AIDS drugs. Researchers, they said, couldn't even give AIDS to a chimp.

Numbers Infected

Another factual uncertainty in 1988 concerned how many Americans were HIV positive. United States Public Health Service projections of 1.1 to 1.5 million HIV positives assumed America contained 750,000 IV-drug users, 2.5 million "exclusively" gay men, and 5 to 10 million men who occasionally engaged in gay sex. These projections were based on Kinsey's data about homosexuality from the 1950s. His data had not been repeated or confirmed for the 1980s. Gays tended to overestimate the number of men who partook of gay sex, whereas others tended to underestimate such numbers. How many people were factually at risk depended on how many gay men (and how many "occasionally" gay men) were out there. Such projections were controversial. Masters and Johnson projected 3 million American HIV positives.

Personal biases also affected projections about how many HIV-positive people would develop AIDS. From 1986 to 1987, estimates were raised from 10 percent to 30 to 50 percent of HIV-positive people would develop AIDS. The dark view among researchers and fundamentalists was that *all* HIV-positive people would develop AIDS and that no cure would be found to save those infected. This Inevitability View—that all HIV positives would develop ARC, then AIDS, and then die—was often repeated as hard fact. As of 1988, it was not.

It was sad to point out that some of these dark projections were desired by some officials who had been elevated to new status. A mighty evil needed mighty warriors to fight against it.

Some people also wanted all gays and IV-drug users to die, if not desiring a gay-free and IV-drug-free America, then at least saying to themselves, "Well, it wouldn't be that bad if they all died." These desires may have influenced people to make projections about deaths from AIDS unwarranted by the evidence.

Prevention

If background assumptions influenced how people viewed facts about the transmission of HIV, it was to be expected that such assumptions would create even more disagreement about how to prevent the spread of HIV. The intense campaigns to eradicate syphilis in America between the turn of the twentieth century and World War II are instructive. Reformers then could not agree about whether the enemy was sin or syphilis. So what happened?

When latex condoms could be widely distributed, reformers split over whether to give them out. Those who opposed sex with prostitutes and nonmarital sex also opposed giving out condoms. Those who simply opposed syphilis favored giving out condoms. During the Second World War, the same crusaders divided over giving out penicillin to treat syphilis for similar reasons. However, generals overruled, and penicillin became standard treatment.

Today, similar battles about condoms rage. The Catholic Church opposes giving out condoms, because doing so will not, it says, in fact stop the spread of HIV. Monsignor Edward Clark of

St. John's University said in December 1987 on *Crossfire* that giving out condoms to HIV-infected people means that their partners will get HIV in September rather than April. Catholic bishops divided bitterly in 1988 over giving out condoms.

Similar battles were waged over giving out sterile needles to IV-drug users in New York City. Even though similar programs had been successful in England and Holland, and even though HIV spread at alarming rates in New York City through intravenous-drug sharing, politicians in New York opposed the program.

In these programs to distribute condoms and sterile needles, as well as programs designed to educate gay men and IV-drug users about safe practices, crusaders revealed opposing evaluative background assumptions that distorted views about AIDS facts. At bottom, the old division still existed between those against the newer forms of sin and those against a new lethal microbe.

Conclusion

No fact about AIDS was considered in a human vacuum. No fact was immune from critique about personal biases that may have distorted its real support. By 1988, just seven years after the discovery of AIDS, it is revealing to look back at all the factual mistakes already made by trusted officials. Indeed, there have been so many mistakes and corrections about facts about AIDS that the credibility of many public health officials has been jeopardized.

Facts about AIDS followed the classic pattern of an infectious, lethal disease with stigmatized victims. Every new factual claim about AIDS was resisted by some and welcomed by others. For those at risk for AIDS, or for those deeply against AIDS-associated behaviors, the stakes were high.

In the end, the politicization of facts about the course of AIDS hurt people with AIDS: it allowed HIV-positive people to continue to deny their risk, to avoid early testing and treatment, and to omit precautions for transmitting HIV. Such politicization came from both sides of the political spectrum. Given the intense dread of a mysterious new lethal disease associated with deviant sexuality, perhaps such politicization was inevitable in the general public. Within medicine, and especially on the part of Robert Gallo, it was inexcusable, for such behavior cost some people an early death.

27

Don't Fear the Human Genome Project

The ethics of in vitro fertilization caused quite a bit of needless concern during the 1970s. Now a similar controversy is unfolding about the Human Genome Project. Within a very few years, couples may be able to obtain much more information about whether they carry the genes for lethal diseases. Surprisingly, what has been written to date about this project has been largely cautionary or even hostile. Nonetheless, this project promises to be one of the most important developments in the long history of human reproduction.

Consider Porter Colley, a sixty-four-year-old Massachusetts woman who has suffered all her life from neurofibromatosis, or "elephant man's disease." "For all these years," she said, "I saw no research and therefore had no hope for a cure." Now she is optimistic that the Human Genome Project will spur genetic therapy for, if not her, then others with her disease.

The clock began ticking in October 1990 for the Human Genome Project, whose $3 billion cost makes it one of the most

---

Originally published as "Ethics and Genetics," *Birmingham News,* 14 June 1994, C1.

significant single projects in scientific history. Already, this project has helped create seemingly monthly breakthroughs in genetics. Exactly what is this project and how will it help parents have more choices?

The DNA inside each human cell contains 3 billion base-pairs of nucleic acids. The Genome Project will produce a map of those. Think of a person's DNA as the American interstate highway system. On it, there will be 30,000 "cities" and "towns." These are the genes transmitting heredity.

The largest gene for disease is the muscular dystrophy gene, composed of 2 million base-pairs—this is a "city" on our "interstate." The globulin and insulin genes are mere "towns" at about 1,000 base-pairs each. At present, we can spot the large cities and some towns, but because mapping is only beginning, we don't know the location of the very small genes.

One goal of the project likely will be met during the 1990s: identifying the genes causing the major known hereditary diseases. Geneticists then will need to discover what role unknown genes play in causing disease, whether a gene influences another gene to cause disease, and how environmental agents interact with genetic material to cause disease.

Experimental gene therapy has already been used for malignant melanoma, and the real payoff in the future will be more of these new genetic therapies, such as for neurofibromatosis. This isn't what the misleading phrase "genetic engineering" implies, where the change in genetic material would be transmitted to children. Indeed, some genetic therapies, such as those delivered by inhaling an aerosol spray, seem much like today's drug therapies.

It is true that the benefits promised from the Genome Project may have been exaggerated and that after the project is done, we will still have a long way to go. Nevertheless, to fly to the stars, it is necessary first to walk on the moon.

As recently as 1987, top scientist Gary Nabel of the Howard Hughes Medical Institute doubted that genetic therapy would ever work or that he personally would ever try it. But, a few weeks ago he said, "I've just been astounded at what has happened in this field in the last couple of years." Now he leads the team trying genetic therapy on malignant melanoma.

Because in vitro fertilization removes several eggs from the ovary and then fertilizes them with sperm in a Petri dish, a cell from each resulting pre-embryo may be checked for genetic defects. Any with a known defect need not be implanted in a woman's uterus.

In unassisted, normal reproduction, about 40 percent of pre-embryos do not become embryos because they do not implant on the uterine wall. Tests on these show they may have genetic diseases. Hence, nature itself now eliminates many defective pre-embryos.

To the dread of some critics, more and more techniques become practical each year that increase choices about reproduction. Amniocentesis initially could only be done late in the second trimester of fetal development, but now it can be done around fourteen to sixteen weeks. Another technique, chorionic villus sampling, can be done at eight to nine weeks. (But both techniques may cause a slight increase in risk of miscarriage.) The earlier in fetal development a diagnosis is made, the more agreement exists over the permissibility of selective termination. Many people believe a fetus close to birth is much the same as a newborn baby with a right to life, but most also believe that pre-embryos are just microscopic clumps and have no such rights.

The value of maps of DNA strands for prospective parents is immense and the Human Genome Project will help. It would be as if you are planning a trip across America and you know that there are certain towns and cities with great, unknown dangers. There soon will be reports available on most dangerous places; wouldn't you want to get them?

Unfortunately, there are people who have organized to deny them. According to a recent profile in *Contemporary Authors,* alarmist Jeremy Rifkin has allied with Jerry Falwell, other Christian ministers, and some rabbis to oppose increased genetic choices in families.

Another group of skeptics voiced their concerns in a special 1991 issue on the Genome Project of the *American Journal of Law and Medicine*. Epidemiologist Abby Lippman there worries that women will be manipulated by society to seek genetic testing during pregnancy. "In today's Western world," she writes, "biomedical and political systems largely define health and disease, as well as normality and abnormality." Sociologist Dorothy Nelkin also warns in this journal that information developed by the Genome Project could make people think that differences in the way children learn in schools are genetic, not social and cultural.

The topic of letting parents choose not to have babies with genetic disease has been so sensationalized in Germany by disability advocates that some famous medical ethicists have been prevented from speaking.

Rather than increasing dangerous knowledge and choice, some say, we should leave things as they are.

Because so much of the essence of these debates is philosophical at the most personal level, it is helpful to pause to think about the most fundamental views about having babies.

Such views broadly fall into two kinds: choice ethics, which increase the range of choices available to prospective parents, and fatalistic ethics, which urge parents to accept things as they have always been.

Fatalists accept random reproduction; choice ethics advocates deliberate, chosen reproduction. Fatalists see primitive, random reproduction as natural and most human. Choice ethicists see chosen reproduction as the most natural and human because it stems from what is highest and best in us: our rationality, compassion, free will, and knowledge.

Surprisingly, many groups on the political left and right endorse the fatalistic view. It is no surprise that social conservatives such as Falwell oppose new choices, but what is surprising is how many women authors pose as feminists to join this opposition. You would think that creating more choices about reproduction for couples was a male conspiracy.

Joseph Fletcher, an Episcopalian priest and moral theologian, is one with an opposing viewpoint. He attacked fatalism in the 1970s:

> It is depressing, not comforting, to realize that most people are accidents. Their conception was at best uninformed, at worst unwanted. There are those who are bemused and befuddled by fatalist mystique about nature with a capital "N" that they want us to accept passively whatever comes along. Talk of "not tinkering" and "not playing God" and snide remarks to "artificial" and "technological" policies is a vote against both human (and humaneness). (*Ethics of Genetic Control: Ending Reproductive Roulette,* 1974)

Of the billions upon billions of people who have been born on Earth, only a minority have been deliberately conceived. It is only within the past fourteen years that infertile couples have had the chance to have their own babies. Clinical medical genetics has only been systematically studied for three decades. It will, I think, take a few centuries for our thinking about procreative ethics to catch up.

Fletcher questioned whether ethics should only be conceived as Thou Shalt Nots. Instead, he thought the standard of right and wrong concerned something as simple as whether it helped or harmed people, whether it was forced on couples or freely chosen.

Is there anything wrong with a desire to avoid children with genetic disease? Surely not. Many couples who have children are at some risk for cystic fibrosis—the gene is carried by 1 in 22 Caucasians. After learning how teenagers with cystic fibrosis suffocate to death, suppose a couple chooses to test their embryos for this disease. What is awful about this? Or dangerous?

What must never be forgotten amid all the overly generalized discussion is the evil of genetic disease and its immense power to destroy.

A sailor with the dominant gene for Huntington's disease jumped ship on the coast of Venezuela in the 1800s and by 1981, hundreds of his descendants were at risk or had the terrible disease itself. Many other hundreds had died of the disease.

Seeing such evil, who would still prefer a fatalistic ethics, where such a gene is allowed to continue to maim and kill?

28

Children's Dissent to Research: A Minor Matter?

Recently one hears more and more stories of parents coercing kids with cancer into research protocols, subjecting them to invasive procedures and misery for a small chance of a cure. Often it is the parents who cannot accept death, not the adolescent, yet it is the adolescent-subject who must suffer the chemotherapy, radiation, and surgery.

*Non*therapeutic research involving minors below the age of eighteen presents special problems for institutional review boards (IRBs). Therapeutic procedures involving minors are relatively easy to justify, and parents often overrule children's dissent to therapy. But should parents similarly be able to overrule children's dissent to nontherapeutic research? Should child-subjects be told about participation in nontherapeutic research? If they are told, should researchers obtain their agreement to participate?

It's good to distinguish here between a child's *consent, assent,* and *dissent. Consent,* short for informed consent, is the legally required agreement of a subject to participate in research. *Assent* is neither

---

Originally published in a longer form as "Children's Dissent to Research: A Minor Matter?" *IRB: A Review of Human Subjects Research* 2, no. 10 (December 1980): 1–4.

legally binding nor legally required: it is a moral requirement to acquire the closest approximation of consent one can achieve within the child's capacity to understand. Finally, *dissent* is more than lack of assent or consent: it is active disagreement.

Legally, therapeutic procedures may be performed on children with the consent of the parents. Laws do not require the child's assent and the parents may override the child's dissent. Such proxy consent is generally regarded as adequate authorization for the performance on children of therapeutic procedures having "investigational" or "nonvalidated" status. However, the law is much less clear about the authority of a parent to consent to a child's participation in nontherapeutic research, especially if the research presents risk of harm to the child.

Four philosophical views on children's participation in research may be contrasted. First, the *Little Adult* view claims that the rights of adults should be extended to children insofar as possible. Self-assertion, independence, responsibility, and autonomy should be encouraged in children. Some people believe that denying children such autonomy constitutes child abuse.

Some child psychologists argue that minors fifteen years or older can understand as much about research as adults. Adolescents ages fourteen or fifteen generally have acquired assertive skills and are more easily able to dissent (inchoate autonomy during such years is often misleadingly called rebellion). When a child under fourteen years dissents, an unusual event, he or she is either unusually intelligent, unusually terrified, or unusually hostile. Any one of these possibilities precludes forcing a child to participate.

In this view, not recognizing the dissent of minors fifteen years and above is hard to justify according to psychological evidence. In this view, a child's assent to participation should be mandatory and such assent should be determinative except in life-or-death decisions. Advocates of this view put the burden of proof on anyone who would overrule a child's refusal to consent or dissent to research.

Second, the *Ignorance* view claims that few adults can give real informed consent to sophisticated medical and behavioral research. As former editor of the *New England Journal of Medicine* Franz Ingelfinger once wrote, most adult subjects give "informed (but uneducated) consent." Since normal adults do not really understand, we know too that minors don't understand. But since we allow ignorant adults to participate in research, why shouldn't we allow ignorant children? As University of Wisconsin pediatrics professor Norman Fost says, "If one acknowledges that decisions to use adults are ultimately as 'coerced' as decisions to use children, and if one concludes that the present system is justified for adults, there is no reason to exclude children from the same benefits."

Third, the *Incompatibility* view claims that nontherapeutic research on children is simply never justified. One who argued this was Paul Ramsey, a conservative Christian theologian, who argued that research on a child can only be justified by intentions to benefit a child directly and never by intention to benefit other children. By definition, nontherapeutic research is not intended to directly benefit a child, therefore such research can never be morally justified.

The Incompatibility view recognizes that it may be evil at times not to do research on children. However (and perhaps inconsistently), it emphasizes that research on children is still an evil. What seems to be meant is that overall benefits to society are never such that the evil of using children for research comes out good overall. Also, some people fear a slippery slope whereby in legitimizing any nontherapeutic research on children, more and more such research will occur.

Fourth, the *Good-of-Others* view holds that it is important to understand that most research involving children is intended to benefit other children, albeit children who may not yet have been born or children who have no relation to the subjects. It is misleading to couch the conflict as one between adult researchers and child subjects; rather, it is a conflict between a minor irritation for a few child subjects and a possible very large good for many children. In this utilitarian view, a small risk to any particular child is justified through its expected contribution to the good of most children.

Among these four perspectives, there are two sets of polarized extremes. The Good-of-Others view justifies any well-designed research with a good risk/benefit ratio. In contrast, the Incompatibility view sees all research on children as evil. The second set of contrasts reflects opposed judgments about the importance of a child's understanding. The Little Adult view attributes great value to a child's assent, whereas the Ignorance view attributes very little value to it.

The Ignorance view is correct in that most adults understand little about research. Nevertheless, it has been demonstrated that adult subjects could be brought to a level of understanding about a particular research project better than the average professor teaching in medical school, far better than the average physician, and far, far better than the average medical student. Certainly such understanding is also possible with minors, especially those in their teenage years. Moreover, even though most people may understand little and have no real freedom to dissent, the dissent of those who are free and who do understand is still very important.

The Incompatibility view is peculiar in one way. One possible interpretation of its dictum that "it is evil to do research on children" is to hold that it is *always* wrong to do research on children. But to admit there are times when it is evil not to do research is equivalent to saying that it is right and good to do research in some cases.

Ramsey might reply, "No, it's not good. In this case, the evil of not doing research outweighs the evil of allowing research on children. But it's still evil. It's just a matter of creating the least evil." To which the reply is: yes, of course, one can define morally right either as creating the most good or preventing the most evil. They come to the same. How one describes the procedures here seems merely semantic.

Ramsey might retort that there is a difference between saying "In some cases research is right, in others, it is not" and "In some cases not doing research creates more evil than the evil of doing research." By use of "evil," he wanted to burden researchers with guilt, and even more so, he implied that this use was a good way for researchers to approach doing *nontherapeutic* research on children.

The Little Adult view is most plausible with articulate, well-educated children who have been successfully trained to assert themselves. Moreover, such children usually recognize themselves to be potential equals of adults in a pluralistic world. Few children obviously fit this view.

Nevertheless, some do. Moreover, some such children are extraordinarily terrified of research, tests, or hospitals. It may be impossible for adults to understand such a child's terror and since such children will be rare, there is no reason for not considering their dissent to be final. Other children may be unusually hostile to the researcher, perhaps out of a general hostility toward adults, teachers, or physicians.

Again, the rarity of such extreme hostility and the degree of reluctance to participate expressed by such children indicate cases where dissent should be accepted. Moreover, the very intelligent child may understand something about himself or herself, about the research, about his or her parents' motivation for agreeing to participate, or even about the researcher, and consequently decide not to participate. Since we recognize similar reasons among adults for declining to participate, we should also recognize such reasons among minors, especially in the teenage years, lest our children's hospitals become torture chambers for some of our most vulnerable patients.

29

Organ Donation Can Kill You

Newspapers recently described the gruesome death of Albany, New York, newspaper reporter Mike Hurewitz at Mount Sinai Hospital in New York City after he donated part of his liver to his brother. Similar terrifying stories about living organ donors are starting to appear, such as that of Barbara Tarrant, a sixty-nine-year-old North Carolina woman who donated a kidney to her mentally retarded son and who wound up paralyzed on her left side and without coherent speech. What's going on here?

A clue exists in a full-page ad for Mount Sinai that I saw in the national edition of the *New York Times* on December 5, 2001. The ad boasts that this hospital is "at the forefront of living-donor transplantation" and does more living-donor transplants "than any other hospital in the country."

I clipped this ad months ago because I thought it shocking to so blatantly encourage such dubious behavior. Widely regarded as heroic in the popular media, living-donor transplantation carries

---

Originally published as "Organ Donors Risk Health," *Los Angeles Times*, 9 April 2002, B13.

real dangers. In 2002 in North America, and for the first time, living donors surpassed brain-dead patients as sources of organs for transplant, yet no one really knows how many people have died from such surgeries and living-donor transplant centers certainly aren't going to tell us.

For many decades, a bright ethical line existed in transplant surgery of "First, do no harm," which meant (among other things), "Even to help another, never render a healthy person dead." A kidney transplant from one identical child-twin to the other first crossed that line in 1954, but that was considered an anomaly because of the perfect match.

Then in the mid-1990s, the dam burst and the flood began. In 1993 Nilda Rodriquez gave one-quarter of her liver to her desperate granddaughter and James and Barbara Sewell each gave part of a lung to their daughter who was dying of cystic fibrosis. By 1997, as the practice became more accepted, California surgeon Vaughn Starnes had taken lobes from sixty-six donors for thirty-seven recipients, a practice one commentator called "ethically problematic."

By September 1999, the *American Medical News* had mentioned the first "confirmed" although "not officially described" death from adult-to-adult liver donation and also guessed that two to three adults had died from donating parts of their organs to children. No one knows today's real figures because they are not by law reportable.

Many troubling ethical questions haunt this trend. If it becomes a norm, how can a parent *not* volunteer to donate? Being a hero becomes a duty. Second, given the heroic mantle wrapped around healthy-donor transplantation, can real informed consent be obtained? Barbara Tarrant seems to have thought few problems existed for a seventy-year-old woman undergoing such surgery. She paid a high price. Third, do the enormous costs of such operations (a liver transplant costs a quarter of a million dollars) and their likely harm to donors justify expansion of this trend?

The answer to this question needs a context and it is this: if there were no alternative to living-donor transplantation, it might still be justified. After all, people die without receiving these transplants, and about four thousand Americans die every year waiting for organ transplants.

Fortunately, there is another way: we could start paying people for the organs of *deceased* relatives. More technically, states could allow "rewarded cadaveric donation," such that families who agreed to donate would get, say, a thousand dollars toward a funeral. Tens of thousands of additional usable organs would now be available to save lives. Why not give this idea a try? It might just work.

True, commercialization is a risk, and we certainly don't want a market for organs between living donors. On the other hand, we can encourage the kind of donation that is already legal for brain-dead patients. And the question is not whether any risk of harm exists from commercialization (it does), but whether such risk justifies the sacrifice of so many thousands of dying patients. I don't think it does.

30

Big Brother Is Watching: The Ethics of Cybermedicine

In Thomas Harris's new novel *Hannibal*, FBI agent Clarice Starling finds her man by tracking down cash purchases of the very expensive, exotic items that Dr. Hannibal Lecter prefers. She uses data aggregation, which collects information generally about people, and which can then be analyzed to reveal specific individuals and their personal habits, health, or histories.

Previously, only law enforcement officials could do this, but today, it is not Big Brother who poses the danger but instead start-up commercial enterprises. Such businesses want to sell data about you to marketers and others who desire such information. In general, information is power, and in particular, information about you is money.

How does this work? Suppose you have twenty fields of data and you know that information about a person is in some of them, but you don't know his name, Social Security number, or medical history. However, you can target this person by using two or three facts, such as sex, age, home address, e-mail address,

---

Originally published (with Jason Lott) as "They Really Are Watching," *Birmingham News*, 4 July 1999, C1.

occupation, or race. With surprisingly few facts, a targeted person can be identified.

There should be a new field in ethics that analyzes ethical and unethical frameworks for collecting and analyzing information. Broadly speaking, ethical frameworks protect citizens, whereas unethical frameworks allow others to harm them. Especially in medical ethics, we are already very late in understanding the uses and purposes of such informational frameworks.

One researcher found that 80 percent of people can be identified through data aggregation by using only the patient's birth date, ZIP code, and gender (*Proceedings of the American Medical Informatics Association* [Fall 1997]: 51–55). In a remarkable coup, this Massachusetts researcher uncovered the supposedly confidential medical records of William Weld, then governor of Massachusetts.

Confidential, sensitive personal information is surprisingly easy to discover via the Internet. Many web sites keep a record of the addresses of computers making inquiries, which can be linked to names of patients. If a portal asks for a person's e-mail address, this makes it much easier to identify a person and his medical records. Some inquiries immediately put one in special groups: If a person makes an inquiry about how to test for HIV or where to buy an at-home HIV testing kit, his inquiry may put him in the group of people engaging in at-risk behaviors.

A large number of American companies self-insure their employees, meaning they create their own pool of money to cover the medical costs of their workers. If some of these companies could hire the same quality of employees without having any workers with serious medical problems, they would make more money.

There are private companies such as Medical Information Systems in Boston that specialize in selling information about individuals to companies that sell life insurance. Other companies are getting into the business of selling information about which applicants should be rejected for medical coverage.

Because of this, the average person needs to be aware that when he fills out survey information for a magazine (fill out this card and we'll enter you in our sweepstakes for a free trip to Hawaii!), he is voluntarily giving out information that may be later sold, with his name attached, to people who may not have his best interests at heart. Even health fairs, where cholesterol and blood pressure are checked, have been used to collect data on individuals.

The new danger is that using the Internet can reveal personal information about you to such companies, and in a way that allows companies to find and sell such information almost instantaneously. Understanding the Internet's role in these informational frameworks has to be part of this new field of ethical investigation.

Christine Varney, commissioner of the Federal Trade Commission, is well aware of the danger. In a recent interview with *PC World* magazine, Varney said, "I think consumers are not aware enough of the potential to aggregate, disseminate data about their preferences and their interests. That when you go online, you're in an entirely different realm concerning data aggregation and data collection."

Some sites want you to use them as portals to free online chat groups or to discussion groups for people with a specific medical problem, such as breast cancer or schizophrenia. Although they may promise confidentiality on the front end, if you click several pages back through the fine print, you will find that they practice data aggregation. One way to look for companies practicing data aggregation is to look for sites that don't list commercial advertisements on their web page. The money to keep the web site has to come from somewhere, and indeed almost all of these sites explicitly state in their fine print that they sell aggregated data in anonymous form to other businesses and corporations.

Often the aggregated data from these medical sites is bought and integrated into a larger pool of data. Companies such as Metromail, First Data Solutions, and Axciom maintain information on more than 140 million Americans and 90 million households. Considering that more than 92 million Americans over the age of sixteen use the Internet (half to access health sites), these numbers are not that surprising.

Once in the larger pool of data, information about individuals is passed along to large businesses, commercial insurance companies, credit bureaus, and, yes, even the federal government. Recognizing the danger is half the battle, although the battle may get much worse. Varney predicts computer programs that observe habits of online consumers by tracking what sites they click on. A possible online database of consumer habits combined with the present online database of consumer information would form a volatile combination, empowering insurance, credit, and private investigatory agencies with personal knowledge about almost every adult American.

Can anything be done to prevent misuse of data? Varney and other government officials think that the problem will solve itself through voluntary standards enacted by the free market. Others, such as the Electronic Privacy Information Center, believe that the voluntary standards to protect consumer privacy have previously failed miserably and will continue to do so in the future.

The American Civil Liberties Union of Wisconsin is worried that the state of Wisconsin is allowing companies to track deadbeat dads by using motor vehicle records, registries, directories of new hires, medical data, claims for unemployment compensation, lottery vendors, lists of winners of lotteries, and tax records. While almost everyone would accept this goal as worthwhile, the ACLU worries about broader uses of the same data for less worthy goals.

Ultimately, we consumers and patients can help decide our own fate. Although recent polls show that concern for privacy ranks high among users of the Internet, little, if any, action has been taken to protect privacy from data aggregation. If we don't make our fears known to the government, it will be unlikely to act to prevent such covert abuses.

31

How to Get AIDS Drugs for Africans

In the fourteenth century, the Black Death stalked Europe and killed a third of its people. At the time, Europeans did not know that the disease is caused by the bacillus *yersina pestis* and that it has both a sudden, pneumonic form and a flea-borne form, bubonic plague. Even if they had understood these facts, they had no medical treatments.

Today another kind of Black Death stalks the world and it is called AIDS. As many as one in three Africans will die of it, especially Africa's future: its children, young women, and young men. Black congressman Ron Dellums (D-Calif.) says that AIDS may crush Africa more effectively than European colonialism did.

Fortunately, we now have weapons against AIDS that were not available against the Black Death. But there is a problem. The main weapons—protease inhibitors and zidovudine—are extremely expensive. HIV-infected patients without medical coverage pay as much as $1,000 a month for these drugs. The steep cost of these drugs puts them out of reach of most African patients.

Originally published as "Balance Drug Research with Saving Lives," *Birmingham News*, 23 January 2000, C1.

The drugs do not cost that much to make, but drug companies charge higher prices to help them recover costs of research and development as well as to give stockholders a return on investment. Americans want new and better drugs in the future, which requires more research and development on the part of drug companies, which need sustained profits to encourage the investors who buy their stocks.

At the same time, not everything in life is about the future health of Americans or profits for drug companies. The fact that 40 million Africans will likely die for lack of cheap drugs must weigh heavily in any moral judgment. Doctors Without Borders recently introduced into Kenya a cheap version of the AIDS drug fluconazole and in doing so paid drug companies no royalties at all.

African nations want the right to manufacture such drugs in their own countries and do not want to pay normal royalties to American and European pharmaceutical companies. U.S. companies have been fighting this right, wanting to maximize profits and fearing that cheap African drugs would find their way back to America. They have been lobbying Congress to inflict trade sanctions on African countries that violate U.S. patents on AIDS drugs.

This fight affected Al Gore's campaign for president. Initially, Gore sided with the drug companies, but after ACT-UP activists dogged his campaign (and perhaps after seeing the moral light), Gore reversed himself.

There is a workable solution to this dilemma, a compromise known as "parallel importing" and "mandatory licensing." This prevents a drug company with a valuable patent drug from withholding it in hopes of extracting exorbitant fees from those who need it. The South African government in 1997 enacted just such an act for AIDS drugs, giving foreign pharmaceutical companies small royalties but not nearly what they wanted.

I believe we should allow the governments of African countries to manufacture AIDS drugs cheaply. This is the best chance of life for the next generation in Africa. Even then, many will either not be able to afford the drugs or not be able to successfully take them. But with cheap drugs, at least they have a chance.

Although a worldwide pool of cheap AIDS drugs poses some threat to the profits of drug companies, this threat is not worth the sacrifice of 40 million African lives. Former vice president Gore's present stance that we should not retaliate against African countries that make their own AIDS drugs is correct. Besides, drug companies are good at convincing North Americans that brand-name drugs are better than generic drugs, and Americans will still want the newest and best drugs, ones that will not necessarily be available in Africa by mandatory licensing. Nor will most Americans have access to black market drugs smuggled in from Africa.

There is also a lesson here for us about patents on genes. The U.S. Patent Office unwisely allowed scientists such as Celera's Craig Venter to patent segments of human genes. This was done even though normal patent conditions require that the thing being patented *not* be a "thing existing in nature" and that the patent must be on something with a proven, useful purpose. Celera was allowed to patent things existing in nature and for which it could not prove any useful purpose (when it applied for its patents, it didn't know which genes governed which functions, nor does it know this now).

This mistake could have serious medical consequences. To prove someone does not have a disease, many genetic tests will need to test for every genetic mutation. Also, the cheapest tests will be multiplex tests, where one test searches for many genetic diseases. A company holding a patent on one genetic segment and holding out for very high licensing fees will be able to block a definitive diagnosis (or block a cheap multiplex test) unless it is compelled to make it available by mandatory licensing.

So both in our own country with genetics and in Africa with AIDS, mandatory licensing of key, patented genes and drugs is needed to balance the incentives of profits for research against the needs of public health.

32

Indigenous Peoples Deserve Profits from Drugs from Their Lands

Several years ago, a member of Earth First! gave a dramatic talk to a standing room–only audience at my university. Among his many reasons for saving the rain forest was the promise of new drugs that might cure age-old human diseases. "The cure for cancer and heart disease may well lie deep in the fungal floor of the Amazon," he said breathlessly.

A year or two later, another environmentalist enlarged on the same point. Mark Plotkin, an ethnobotanist, gave a rousing talk with beautiful color slides showing the great variety, utility, and beauty of rain forest flora and fauna. Again, there was an SRO crowd, and again the speaker made the point about getting drugs to benefit the folks back home.

Plotkin explained how he worked with local shamans in Central America and South America to gain their trust and to learn their secrets: He told how he had watched dramatic cures of headache and swelling and how, in at least one case, he may have stumbled upon a drug hitherto unknown in the modern medical world.

Originally published as "Those Living in Rain Forest Should Profit from Its Secrets," *Birmingham News*, 5 November 1995, C7.

A few weeks after Plotkin's talk, the *Wall Street Journal* carried a front-page article on his work and his new company, Shaman Pharmaceuticals, a for-profit company designed to harvest promising drugs from the rain forest and test them in clinical trials in North America and Europe.

I relate this history of recent talks about the rain forest and environment because it bears an important relation to a little-noticed event at the Fourth World Women's Conference in Beijing in September. While researching material for a senior seminar on bioethics and environmental ethics that I'm now teaching, I stumbled on a press release on the Internet that I had never seen in any American paper. Outside the official talks of this conference, a small group of Latin American women asked that their peoples be allowed to share in the profits of international pharmaceutical companies. They claimed that these profits were gained by searching their ancestral medicinal wisdom for drugs to test, often without the consent of the leaders of the indigenous peoples.

The request of these indigenous women was so just, their reasoning so lucid, that I was ashamed that the idea had never before occurred to me. Yet in all the talks that I've heard on environmental ethics and the rain forest, not one speaker has ever mentioned that the shamans' medicinal knowledge might rightfully belong to their indigenous peoples, not the first North American or European company that can pry the information out of them. Certainly we should not repeat, in terms of medicinal rights, what we did to the land of the indigenous peoples of North and South America.

A woman named Nuria Pacari, of the Quichua ethnic group in Ecuador, said that some existing international accords could be enforced to give indigenous people an actual patent on their ancestors' knowledge of herbs, plants, and medicines. Obviously, such enforcement will be resisted by large pharmaceutical companies, which typically take an all-or-nothing approach: We get all the profits or we won't sponsor the drug for clinical trials.

Pacari is a member of the board of directors of the Inter-American Human Rights Institute and claims that if native indigenous peoples were granted patents to their folk medicines, they would become self-sufficient. Such knowledge may be the last marketable treasure of these peoples.

Pacari charged that the official governments of these peoples were in cahoots with international pharmaceutical companies. She said they denied indigenous peoples their property rights and marginalized women who had such knowledge. Our people, she said, suffer "from being poor, indigenous and women." Pacari asked that monies from patent rights be granted to the peoples themselves, and not to their national governments, because the governments would never return the monies to the indigenous peoples. Above all, she said, Latin American indigenous peoples do not want to be "treated like exotic elements, or decorative museum pieces or objects to be studied, but as people with valuable thoughts, feelings and philosophies."

Although some of Pacari's claims are extreme, it seems to me that we do indeed owe indigenous peoples some return if we want to go in and (literally) rip out their pharmaceutical knowledge for our own benefit.

Aspirin came from acetylsalicylic acid in willow bark, and cyclosporin (which prevents the body's rejection of organ transplants) from a fungus. The drug penicillin is obtained from the molds of the genus *Penicillium.* Who knows how many other wonderful drugs are in the rain forest? Who knows? Well, the native shaman and medicine woman, that's who. But why should they cooperate with Mark Plotkin and others for his profit, our benefit, and their own demise? Why indeed.

We should guarantee them a share in the profits. It's the right thing to do.

33

Norman Borlaug: He Fed a Billion People but You Don't Know His Name

Norman Borlaug does not look like a hero, as least not the way Hollywood movies portray one. A typical elderly white male with a rounded face, glasses, and thinning hair, he looks like someone who could be walking around a retirement community.

And yet, in a world that some say lacks real moral heroes, Norman Borlaug has led a life that equals those of Albert Schweitzer and Mother Teresa.

So what has Borlaug done to merit such praise? Led a military raid on Entebbe? Discovered a new kind of drug for arthritis? Adopted a dozen disabled children? The answer: as a result of his life's work, a billion people now exist who otherwise would have starved to death, died of starvation-related diseases, or never have been born.

Thirty years ago, as a young college graduate, Borlaug first directed the Rockefeller Foundation's Mexican wheat program, formed initially to teach Mexican farmers new agricultural ideas. In the small beginnings of the Green Revolution, Borlaug developed

---

Originally published (with Joyce Hsu) as "A Hero for Our Time," *Birmingham News*, 23 July 2000, C1.

dwarf wheat and a technique called shuttle breeding. By shuttling seedlings twice a year between two regions of Mexico a few hundred miles apart, Borlaug was able to develop a variety of wheat that grows well in a range of climates, altitudes, and seasons.

Selectively breeding the dwarf wheat (already naturally resistant to a variety of plant pests and diseases) for semidwarfness forced higher yields. With the plant devoting less energy to growing a tall stalk, more energy went into growing edible grain, doubling and tripling traditional yields.

Borlaug's agricultural approaches benefit people in many ways. His work has fed billions of people in developing nations, created jobs, preserved the environment, and indirectly improved many lives. How has he done this? Well, his approaches to agriculture, which use relatively small plots intensively farmed with chemical fertilizers, do have side effects. Because these crops depend greatly on humans, their highest yields require planning and constant care. Demand for machines that sow and harvest such crops spurs regional industries that make machines. New factories in turn create more jobs. With modern technology helping out, birth rates decrease because farmers need fewer children in the fields. Managing a larger, more productive farm requires knowledge, encouraging parents to have fewer children and to educate existing children. A stabilized population results.

As for the environment, traditional agricultural practices in many developing countries employ slash-and-burn techniques. Such practices destroy more pristine land than Borlaug's high-yield practices, which replenish fields with fertilizers and make the same area produce several times more food. In the past, soils would be depleted after a few seasons and farmers would then cut down more forest for farmland with no increase in production. Paradoxically, Borlaug's high-yield methods actually preserve grasslands, wildlife areas, and rain forests.

With Mexico successfully producing dwarf wheat, it made sense to Borlaug that other countries, such as India, could improve production by using his techniques to grow new varieties of cereal. Environmentalists protested that developing countries should grow their own indigenous crops and grow them using organic methods. Borlaug responded simply that starving people needed food now and indigenous crops grown by organic methods did not yet produce high yields.

From Mexico, Borlaug moved on to Pakistan and India. Malthusian pessimists such as biologist Paul Ehrlich, population ecologist Garrett Hardin, and World Watch Institute head Lester Brown claimed that facts contradicted Borlaug's goals. They claimed that the population explosion would always surpass food production and that the Indian subcontinent would always suffer disastrous famines. These three prophets of doom won the war of influence with the public, which subsequently became fatalistic about famine.

Although constantly criticized by these doomsayers, Borlaug wasted no time bringing a starving Pakistan to self-sufficiency, closely followed just a few years later by India. In the past thirty years, India's population doubled, her crop production tripled, and her economy grew nine times. At one point, despite war and unrest, India even exported cereal grains.

Soon after these successes, Borlaug and his colleagues introduced a high-yield variety of rice throughout most of Asia. But then the doomsayers won. Foundations such as Rockefeller—which had supported Borlaug's work for years—yielded to the protests of environmentalists (especially Greenpeace of Europe) and ceased funding Borlaug's work. Years later, backed by former president Jimmy Carter and funded solely by Japanese multimillionaire Ryoichi Sasakawa, an eighty-four-year-old Borlaug sought to bring his agricultural revolution to Africa. Problems with civil unrest and a lack of infrastructure made success difficult there, but test plots still grew as he predicted.

Contrary to popular belief (and partly as a result of Borlaug's work), the amount of food per capita in the world has actually increased over the past decades. Indeed, most sides of debates about ending famine agree that the world now produces enough food for everybody. Some then argue that the problem is one of distribution: for example, getting food from North America to Africa.

Borlaug disagrees. He thinks famine will only be stopped when poor countries develop their own high-yield crops, use chemical fertilizers and genetically enhanced crops, and nurture regional food economies. The best target for charity is not buying food from rich countries and sending it to poor countries but making poor countries self-sufficient by helping them use high-tech agricultural science.

Amazingly, a large coalition of European and American organizations actively oppose Borlaug's ideas for poor countries with starving peoples. Organizations such as Jeremy Rifkin's Pure Food Campaign see the scientific techniques of modern agriculture as the evil knowledge of international agribusiness. Such organizations want instantaneous, egalitarian land reform combined with organic farming to create self-sufficient, ecotourist-friendly countries. Even though genetically enhanced golden rice (rice containing a bit of carrot) could get vitamin A to African children and prevent thousands from going blind, Greenpeace and the David Suzuki Foundation oppose planting such rice.

Recently, seven academies of science urged the use of high-yield techniques—including the use of genetically enhanced beans, wheat, and rice—to alleviate world hunger. These academies urged us not to focus on the process of adding a desirable trait to an old crop, but on the actual effects of the new crop on people and environments. If we follow their recommendation, we will follow Norman Borlaug's wonderful legacy.

34

Hating Biotechnology: A Tree with Deep Philosophical Roots

Can you be a Green Warrior in a peaceful cause? Can Green Warriors realize too late that they serve a Dark Lord?

Americans such as myself differ from Europeans on destruction of trials of genetically modified (GM) crops. After the September 11 attack on the World Trade Center, our tolerance for acts outside the law has decreased. Not having faced mad cow disease, we Americans may also be more tolerant of GM food.

So from my point of view, dangers to the world's food supply now include not only the traditional floods, drought, and pests but also suspicious activists who destroy field trials of GM plants. In the roughly 150 sites in the United Kingdom and Europe where GM crops have been tested, more than half have been destroyed by misguided activists working for groups such as Greenpeace.

In the United States, radical groups destroyed thirteen such field trials in 1999. One such group, Reclaim the Seeds, destroyed a trial of sugar beets at the University of California at Davis, asserting, "Modern agri-business and genetic mutilation is [sic] a capitalistic machine that must be dismantled."

Originally published in *Philosopher's Magazine* (London, England), June 2002.

In the United States, a spokesperson for the Earth Liberation Front claimed responsibility for burning down a research lab in the University of Washington Center for Horticulture, which grew trees that were genetically enhanced to provide more pulp for paper. The vandals also destroyed one hundred of the three hundred remaining plants of showy stickweed, an endangered plant being nurtured at the lab.

Vandals break the law and hide; conscientious objectors break the law as public protest and suffer the consequences. Most of the attacks so far on GM field trials have been by vandals.

In the most famous attack in England by conscientious objectors, Lord Melchett, the executive director of Greenpeace, and twenty-six Greenpeace members trespassed on a farm in Dereham, Norfolk, in July 2000 and destroyed 2.4 hectares being grown of Bt corn (corn containing the genes of *Bacillus thuringiensis*, a natural pesticide found in soils and also sprayed on organic crops). The sixteen men and ten women claimed that they had a legal right to destroy the crops to protect the environment. As one who reads its literature, I believe that Greenpeace has decided on a priori grounds that the experiment, designed to test whether Bt corn really endangered the environment, should not proceed.

But how can a test, which might prove GM crops to be harmful, be construed as bad? Anti–GM food groups answer that those designing the tests cannot be trusted or that if the tests show safety, it will be the thin edge of the wedge toward accepting more GM crops.

I can't accept this reasoning. If there are real dangers from GM crops to people or the environment, then objective field trials should be carried out. What is it that opponents fear in such trials? That the crops will turn out to be safe and benign to the environment?

Arrested and tried, Lord Melchett and his merry green band welcomed the ensuing publicity, which, of course, they had orchestrated. Despite overwhelming evidence that they had broken the law (they essentially admitted that they had broken the law), they were acquitted.

The ethical thinking behind these actions is that the end justifies the means. But does it? Does destruction of GM food really stand on firm ethical ground? What soil nurtures this thinking? From what historical roots? The answers disturb concerned citizens, both for what passes as reasonable in ecological ethics and for the roots of such thinking in . . . surprise! Nazi ideology.

The intellectual connection between radical environmentalism and Nazi promotion of "blood and soil" *(Blut und Boden)*, the purity of the Fatherland, and organic farming needs further attention. Radical theories of deep ecology promote interests of ecosystems over those of humans, just as Nazism promoted the interests of the Land, the State, and Aryan peoples over the interests of groups of German citizens. For now, we must take care that unthinking environmentalism not degenerate into environmental fascism, with charismatic leaders manipulating followers through emotional appeals.

To understand the roots of today's anti–GM food environmentalism, we must see what has been taught in philosophical courses on environmental ethics and in related, interdisciplinary courses such as environmental sciences and environmental studies. Because of the influence of ecofeminism, non–sentience-discriminating utilitarianism, and Eastern mysticism, nonanthropocentric or biocentric theories now hold sway over gullible students. Philosopher Paul Taylor's influential deep ecology argues for the equal inherent worth of all organisms in the ecosystem, implying that destroying an ant is the moral equivalent to killing a human baby.

Ecofeminists such as India's radical Vandana Shiva see biotechnology as emblematic of white, male, analytic, phallocentric, anti-Gaia values. Shiva wants an India of rural food cooperatives run by women who share ideas and profits and who use cow dung rather than chemical fertilizers. She mistakenly thinks modern women should spend all their time growing food. Biologist Mae Wan-Ho, a former reader in biology at the Open University in London, follows in Shiva's wake, condemning GM plants and biotechnology as bad science, capitalistic, and dangerous to humans.

Philosophers William Devall and George Sessions champion a deep ecology based on Eastern mysticism, claiming that "all things in the biosphere have an equal right to live and blossom and to reach their own individual forms of unfolding and self-realization with the larger Self-realization. The basic intuition is that all organisms and entities in the ecosphere, as parts of the interrelated whole, are equal in intrinsic worth" (*Deep Ecology: Living as If Nature Mattered,* 1985).

Usually only preached to the choir, such nonsense certainly falls apart under scrutiny. A human baby and an ant do not have equal moral worth. It would be wrong to save a thousand ants rather than a human baby. It would be right to kill a thousand ants to save a baby. Any ethics that implies otherwise is stupid and, worse, dangerous.

Many people accept radical environmentalism without thinking through its possible harms to humans. American writer Edward Abbey said he would rather shoot a man than a snake. Let's hope he was just posturing.

Greenpeace and Shiva don't want starving people to get Bt corn, other GM food crops, or Golden Rice to cure blindness caused by vitamin A deficiency. Are they callous or insane? Even if sowing Bt corn destroyed the environment of Ethiopia, if the resulting crop prevented the death of millions from starvation, it would be justified. But no real evidence proves that Bt corn hurts the environment. Are people starving for lack of GM food because Western do-gooders want to keep these environments pure for Western tourists?

One gets the feeling reading the philosophers of Deep Ecology that the best way to protect the planet would be for the human species to commit mass suicide. In that way, Earth would be purified of all human toxins. In that way, the environmental cancer on the biosphere that is humanity would be cured once and for all.

At its heart, Deep Ecology hates the masses and their aversion to backpacking and kayaking in the wilderness. Such environmental elitism has a philosophical pedigree: Jean-Jacques Rousseau, Ralph Waldo Emerson, Henry David Thoreau, and Nazi philosopher Martin Heidegger all touted the values of communing with Nature far away from technology, urbanization, and the hoi polloi.

But an even darker pedigree haunts Deep Ecology, that of National Socialist thought, as Peter Staudenmaier documents (*Ecofascism: Lessons from the German Experience,* 1995). This environmental ideology extolled purity of race, purity of the Fatherland, purity of the German forests, and yes, purity of food. Strict vegetarians, both Hitler and Rudolf Hess championed organic foods and converting biotech farms to organic ones. Nazi intellectuals championed *Blut und Boden,* the mystical connectedness of Germans to their land, crops, and food. German racial theorist Richard Darre said Jews were "weeds."

One night several years ago, Dave Foreman came to my university to hold a rally for Earth First!, which has been suspected of many acts of environmental terrorism. A young all-Aryan crowd greeted Foreman, looking up to him as their demigod. Above all, I sensed that they wanted to join him in The Righteous Cause. They were like Nietzsche's young, golden lions, wild and barely tamed, hungering for purity of action and if not that, well . . . just action.

As were the youth who rioted in Seattle at the World Trade meetings in December 1999, and those who provoked police at the summit meetings in Genoa in 2000, and those who continue to vandalize GM crops around the world. Are they members of democratic organizations who vote on actions and who follow Roberts' Rules of Order? Not at all. More like blunt instruments, hurled at unsuspecting opponents who, unlike these radical youths, obey the law and who think humans matter more than plants.

Of course, environmentalism pervades chic circles. Environmental groups easily raise millions. In America in 1990, environmental groups raised ten times more money than the Republican Party. Does this explain why Greenpeace moved from defending whales and ocean life to destroying field trials that might prove GM crops safe? Did Greenpeace need a big issue to continue to raise funds? After mad cow disease in Europe, did it see GM food as its big chance, regardless of the science? Why did Greenpeace founder Patrick Moore quit in disgust over such antiscientific stands by his organization? In my opinion, Greenpeace conveniently piggybacked on Deep Ecology, using its shallow ideas to fill its own, quite deep, pockets through sensationalistic fund-raising.

Philosopher Mary Midgley once wrote *Evolution as a Religion.* She had a clever idea, for undoubtedly some scientists think that way. But millions more ordinary people embrace Environmentalism as a Religion, and the followers of this new religion are more self-righteous and more well-heeled. Hollywood movie stars and media billionaires such as Ted Turner give millions to groups opposing GM crops such as the David Suzuki Foundation, Earth First!, Earth Liberation Front, and Friends of the Earth. As in the years before Hitler, UnReason here reigns in politically correct robes.

Of course, apologists say, it's only food and no one is going to die over such minor food fights. Begging to disagree, food security in most of the world constitutes life, power, middle-class status, self-sufficiency, economic independence, and freedom from disease. If we want to help the peoples of the world obtain those goods, we should not tolerate food terrorists or their philosophical ilk.

Index

About the Author

Greg Pence has taught bioethics for twenty-five years in both the medical school and philosophy department of the University of Alabama at Birmingham (UAB). After growing up in the suburbs of Washington, D.C., he graduated from William and Mary in 1970, earned a Ph.D. in philosophy from New York University in 1974, and has been married for twenty-five years (to the same woman).

At UAB, Pence won its highest teaching award in 1994, served on its committee reviewing experiments for twenty-two years, and presently both serves on the Hospital Ethics Committee and directs a program for gifted undergraduates preadmitted to medical school.

His text *Classic Cases in Medical Ethics* (McGraw-Hill) will soon go into a fourth edition. Pence also published *Who's Afraid of Human Cloning?* (1998), *Re-Creating Medicine* (2000), and *Designer Food* (2002)—all with Rowman & Littlefield. He is known for his defense of biotechnologies such as cloning and genetically modified food. He is a popular speaker on these topics at American universities.